BOX KITES TO BOEINGS

BOX KITES TO BOEINGS

THE SKY-HIGH HEROES OF AUSSIE AVIATION.

MAL WALDEN

Published by Brolga Publishing Pty Ltd
ABN 46 063 962 443
PO Box 452
Torquay 3228 VIC
Australia

email: markzocchi@brolgapublishing.com.au

ISBN: 9780648612094

Printed in Australia
Cover design by Luke Harris, WorkingType Studio
Typeset by WorkingType Studio

Aerial Timeline

Preface

Australian history of aviation has borne witness to some of the world's most audacious daredevils and dreamers. Their extraordinary exploits would read more like fiction had they not all been so meticulously documented as fact

The Dreams of Flying

'Once you have tasted flight, you will forever walk the earth with your eyes turned skyward, for there you have been, and there you will always long to return'.
Leonardo da Vinci

In November 1783, a mere four years before the arrival of the first settlers in Australia, the world marvelled at the spectacle of 'the world's first manned balloon flight' in Paris.

Not since the ancient Greek legend of Pegasus, the winged horse, or the cautionary tale of Icarus, whose waxed wings melted when he dared to fly too near the sun, had humanity's dream of soaring like birds come so tantalisingly close to reality.

China's discovery of the principles of kite flying dates back to around 400 BC, while the brilliant mind of Leonardo da Vinci explored the realm of flight in the 1480s, although his vision of the Ornithopter flying machine never materialised.

However, in 1783, the French chemistry and physics teacher Jean-François Pilâtre de Rozier etched his name as *'the pioneer of aviation'* by constructing the first hot air balloon capable of carrying passengers.

This historic event unfolded in the presence of King Louis XVI, who suggested that two condemned criminals be onboard, just in case. However, de Rozier believed that the honour of becoming the world's first balloonist should be reserved for someone of higher status. Thus, the Marquis d'Arlandes courageously agreed to accompany him.

After a series of rigorous tests to gain experience in controlling the balloon, they embarked on their first untethered flight in a Montgolfier hot air balloon.

The flight lasted 25 minutes, covering a distance of approximately 5½ miles (around 9 km) and reaching an altitude of 3,000 feet (almost 1,000 metres) before safely returning to Earth.

Regrettably, de Rozier's fame was short-lived. He met his demise when his balloon crashed during an attempt to cross the English Channel, becoming the first recorded fatality resulting from an air crash. Nonetheless, others continued to chase the dream, with Orville and Wilbur Wright's historic powered flight at Kitty Hawk, North Carolina, in 1903, marking a pivotal moment in the history of aviation.

Almost from its days of settlement, nowhere has this desire to fly seen such development, technical firsts and heroes in aviation than in Australia.

This tribute is a testament to those incredible Aussie aviators and their extraordinary flying machines, which have left an indelible mark on our history and the story of flight.

With thanks to the State Libraries of Victoria and New South Wales and acknowledged contributors.

Henri L'Estrange

Balloons and Beyond

THE BALLOON ACCIDENT.

In the records of history, not all our aeronautical pioneers kicked up the dust from their flimsy propellers. Nor did they all seek to defy gravity for purely scientific purposes. The mid-19th century witnessed a series of daring balloon flights in Australia, each with its tale of triumph or misfortune.

The earliest attempted flight, dating back to 1856, was

attributed to M. Pierre Maigre. However, the paying spectators felt so aggrieved at the aeronaut's failure to fly they demonstrated '*equally aggrieved behaviour*' towards Maigre and he was lucky to escape with his life.

In 1858, balloonist William Dean achieved a more successful voyage. His gas-filled balloon, named "The Australasian", covered a distance of 30 kilometres with two passengers on board.

Thomas Gale's first flight in 1869 took a tragic turn when his balloon tore apart, ejecting Gale from his basket causing serious injury.

The year 1878 also marked a series of misfortunes for Rufus Wells during his balloon flights.

The most memorable took place at a balloon launch from Sydney's Domain when one of his would-be aeronauts jumped out, leaving his companion to fend for himself.

Yet, none could match the exploits of the Australian aerialist Henri L'Estrange.

Born in the Melbourne suburb of Fitzroy in 1842. By the age of 30, Henri had become a household name; not only for his dramatic high-wire aerial stunts but rather sadly because of his accidental mishaps.

Calling himself the '*Australian Blondin*' in homage to France's renowned tightrope walker Charles Blondin, Henri's reputation grew along with his audience.

His journey into the public eye began in 1873 as part of the Royal Comet Variety Troupe in Melbourne, where he dazzled alongside his partner, Miss Lulu L'Estrange.

In 1876, the not-so-humble Henri decided against sharing the spotlight on the tightrope and became a solo act quickly gaining a reputation as a fearless aerialist by adding gas ballooning to his repertoire. The idea that people could be lifted from the ground to fly, and then return safely, quickly fired the imagination of growing crowds. There was also the anticipation of an unexpected accident, which was not uncommon in the field of ballooning.

In November 1878, Henri L'Estrange arrived in Sydney amid reports of a successful balloon flight in India. However, during preparations, L'Estrange struggled to fill his balloon with gas. His calculations, based on the quality of Sydney gas, were underestimated by a staggering 700 pounds. This miscalculation caused his balloon to rapidly ascend to unexpected heights, before bursting. With the aid of a parachute, he descended safely but badly shaken. Unfazed, L'Estrange sought permission for a second attempt, which, like the first, ended in failure.

Less than six months later, he faced an even more disastrous attempt in Melbourne, with his balloon named Aurora.

By the late 1880s, Henri's popularity had reached its zenith, aided in no small part by a series of flights. One was a success, another marked Australia's first emergency parachute incident, and the last ended in a massive fireball, causing property damage and a series of injuries.

Despite these setbacks, L'Estrange persevered and in March 1881, he returned to Sydney, hoping to restore some credibility.

Over 10,000 spectators gathered in the Outer Domain to witness the event.

The lift-off began at 9:30 pm, with L'Estrange seated in a loop of rope, reminiscent of his attempt three years prior. Initially, all seemed well as the balloon ascended above the crowd's heads, hovering briefly before heading over Hyde Park. L'Estrange recounted the rest of his journey in a letter to a friend:

"I then got into a westerly current that took me out to sea. I found the escape valve would not act. In sheer desperation, I took the

valve rope in both hands, and it opened with a bang; but in the effort, I had lost my seat in the loop, falling about six feet, and there I was dangling in mid-air, clutching the valve rope, with gas rushing out of the balloon as though she had burst."

Recognising the danger, L'Estrange deployed grappling hooks as the balloon drifted back over the suburbs. Unfortunately, the ropes became tangled, and the hooks were too short. The culmination was a collision with a house and a fiery explosion. Henri managed to leap from the fireball, but the incident left several bystanders burned and the entire suburb illuminated.

After this final failure, Henri L'Estrange decided to return to his original career of tightrope walking. By now with new forms of entertainment, public falls, and a surge of imitators, achieving success was becoming increasingly elusive.

Public benefit events were organised in his honour to help mitigate his financial losses but sadly fell short. He briefly explored setting up amusement rides before ultimately fading from public view, with the last recorded sighting of him in Fitzroy in 1894.

Despite a chequered career as an entertainer, the legacy of Henri L'Estrange endures to this day as one of Australia's early pioneers, defying gravity and taking to the skies; even if those flights were powerless.

The power of flight was just beyond the clouds.

The Aussie Airship Dream

Daredevils and Dreamers:

'White Australia'- First Australian Airship.

The history of flight has a rich legacy filled with eccentric, reckless, and extraordinary daredevils and dreamers.

In 1851 Sydney Doctor William Bland, designed and patented his 'Atmotic Ship', an early steam-driven airship, supposedly capable of flying to London in a week and a half. But it was just a dream and never materialised.

Another dreamer was electrical engineer Alban Roberts. Born in New Zealand before the First World War, Roberts travelled extensively (not in airships), promoting a belief that airships were the future of flight.

It was in America where he first experimented with his dirigible design, attempting to fly from New York to Philadelphia which almost ended in disaster when he crashed into the Atlantic. He miraculously survived but was determined to fulfil his dream and returned to NZ in 1913 where he perfected a remotely piloted dirigible. His aim this time was to demonstrate its ability in warfare by dropping a bomb of 'confetti' on its designated target and while his aim was accurate enough his 'confetti' failed to impress everyone else.

A few months later, he arrived in Sydney where he built his first full-size airship, named 'White Australia'. There's no explanation as to why he named his airship after the government's 'White Australia Policy' at the time.

Although the most obvious explanation would appear to be his desire to impress the Government in the hope it would be considered in the interest of defence.

The first of his three flights, was a tethered test at the Sydney Agricultural Showground on 23 June 1914, in which the envelope filled with hydrogen, rose 80 feet into the air, just as far as the ropes would allow.

His second attempt on 6 July allowed the airship to move a little more freely under its power. However, according to press reports, this was not a particularly impressive feat when part of the flimsy nacelle (platform) beneath the canopy collapsed providing some spectators with an element of excitement.

'It rose 500 ft. above Rushcutters's Bay but when the engine failed, the dirigible descended on nearby tramlines.

Mr Roberts escaped injury when thrown from the platform he was standing on and was left hanging onto the wires before dropping to the ground. Several thousand gathered to watch, forcing tram traffic to be suspended while the dirigible was emptied and rolled up.'

His final public flight was in Melbourne on 3 October and on this occasion, the press described it as almost flawless:

'On the words 'let go', the airship flew over the fence; cleared the nearby telegraph and electric tram wires, and headed for East Melbourne. Under the control of Captain Penfold, he performed several graceful circles in figure eight. However, when the airship left the show grounds and reached an elevation of 2000 ft the engine was stopped, and the airship landed heavily sustaining slight damage.'

There is little or no evidence of any further public flights after that although Roberts' continued to promote the future of his airships.

By 1930 Roberts finally patented a dirigible for the use of aerial advertising, a dream that wouldn't be fully realised for another 70 years with the arrival of commercial blimps.

As the history of flight evolved so too did the design of dirigibles. Blimps of various shapes and sizes were designed to carry passengers in a gondola hanging beneath the balloon, as opposed to airships which were designed to carry people inside the structure itself. Meanwhile, the dreaming continued.

The Imperial Airship Scheme

In 1926 Britain embarked on an ambitious plan to build two massive airships capable of travelling vast distances, most notably to her far-flung colonies such as Australia.

Stanley Bruce, Australia's then-prime minister, was shown a scale model of a proposed 200-metre-long airship that could make the journey to Australia in just 12 days.

In mid-1927, a delegation called the Airship Mission was sent from Britain to investigate potential sites for airship mooring masts and to generally drum up interest in the scheme. Melbourne was chosen as one of those sites due to its unchallenging terrain.

By 1929, Britain had built two massive airships — the R100 and the R101 — which were housed in enormous side-by-side hangars at Cardington, north of London.

The R101 became the largest flying aircraft ever made. At a length of more than 220 metres, it was dubbed "the Titanic of the skies". It was luxurious compared to other aircraft at the time, with 50-passenger cabins, a dining room for 60 people and two promenade decks.

It was hoped the maiden flight of the R101 would be led by a colonial crew of three distinguished Australian airmen.

Lieutenant Commander Noel Atherstone had won the rare distinction of being the first to sink a German submarine from an airship in World War I. But after the war, he retired to the outskirts of Melbourne in Victoria to become a pig farmer. However, the offer to become part of this Imperial Airship Scheme was enough to lure him back to Britain.

Melbourne-born William Pastra who was awarded a military cross for shooting down six enemy aircraft was also chosen, as was another distinguished Australian, Flight Lieutenant Charles

Harman. But at the last moment, Harman decided to retire and his position on the flight deck was awarded to William Pastra.

The Associated Press reported that both men had previously voiced concerns that they "didn't like the idea of 5 million cubic feet of gas above them".

Nevertheless, the R101 departed Britain on October 4, 1930, bound for Karachi.

As the giant R101 crossed the English Channel it was struck by strong and gusty winds near the town of Beauvais, north of Paris.

A split suddenly developed in its outer cover, causing it to drag and then nosedive. Despite the efforts of the crew, the R101 dropped from an altitude of 1,200 feet — about 365 metres — in just over two minutes, to touch the ground on the edge of a forest.

Although the R101 only made relatively gentle contact with the ground, a fire ignited the airship's gasbags and it exploded, reducing the craft to a charred frame.

Forty-eight of the 54 people onboard the airship died — more than were killed in the Hindenburg disaster seven years later.

Among the victims of the crash were Australians Atherstone and Palstra.

Flight Lieutenant William Harman was left to personally deliver the news of the crash to Palstra's widow on his return to Melbourne by ship the following month.

Back in Australia, PM Stanley Bruce lost the general election in 1929, and the Imperial Airship Scheme was abandoned the following year.

Meanwhile, the dream of Airships continues to this very day.

Thanks to State Library Victoria, Tim Callanan ABC, Trove, The Argus and Airminded.com

Lawrence Hargrave

First Flight

December 17, 1903, stands as an iconic date when Wilbur and Orville Wright achieved the first powered flight at Kitty Hawk, USA.

Orville piloted their gasoline-powered biplane, staying aloft for 12 seconds, covering a distance of 120 feet – a historic moment that etched their names in the record books.

But was their achievement inspired from down under?

* * *

Nine years earlier, on a breezy fresh morning in November 1894, Australian explorer and inventor Lawrence Hargrave arrived at Stanwell Park beach situated between Sydney and Wollongong on the NSW coast. With him was his companion and employee, James Swaine, who shared Hargrave's unyielding curiosity and adventurous spirit. Both were convinced that this windswept beach held the potential to usher in a new era of human achievement.

Located, not far from their homes, the picturesque location was already known for its ideal conditions for flying kites. However, Hargrave saw more than just a playground for children's dreams; he envisioned it as the possible birthplace of aviation.

They began by assembling a series of four box kites, each intricately connected to the one above it.

Hargrave, always one to consider safety, tethered his contraption to the ground with piano wire.

A sling seat was attached to the fourth and final kite, alongside an anemometer and clinometer – instruments to measure wind speed, altitude, and the angle of the kite line.

Then with a gust of wind as their catalyst, Hargrave slowly

ascended more than 4.8 meters (16 feet) into the sky. In that breathtaking moment, Lawrence Hargrave etched his name into aviation history as the first person to soar with a stable fixed-wing design.

Hargrave's achievement was not just a product of ingenuity; it was the result of an intense study of nature's flight mechanisms. He had spent countless hours observing the graceful movements of fish and the sinuous slither of snakes, gleaning insights that would prove invaluable in his quest to discover the motions of flight.

While others were experimenting with mono-winged gliders, their efforts often ended in instability and danger as they sought to harness the power of lift. Hargrave's radical wing design defied convention and offered both greater lift and stability than any previous attempts.

His box kite, with its improved lift-to-drag ratio, had finally provided the elusive foundation for flight.

The late 19th century was a time of great excitement in the field of science. "Gentleman scientists" like Hargrave toiled away in their home workshops and sheds, competing in a global race to invent the first self-powered, fixed-wing flying machine. Hargrave had now set the stage for a revolution with a design that captured the attention of inventors worldwide.

One such admirer was Octave Chanute, an American friend and collaborator who shared Hargrave's dreams. Hargrave's pioneering work had already made its way across the ocean, influencing aviation enthusiasts in the United States. Chanute, in an official journal, marvelled at how the skies of eastern USA were "red with Hargrave Kites," as groups of aviators proudly proclaimed themselves "Hargrave disciples."

The French, who believed themselves to be the cradle of aviation also recognised Hargrave's genius. When the first

European aircraft were built, they too used Hargrave-type box kites for their supporting surface. But while Hargrave's influence was being acknowledged far and wide, some sought to downplay his contributions.

American officials were denying his influence on the Wright brothers in a dispute involving politics and patents.

Hargrave was never driven by the pursuit of personal recognition or fortune. He firmly believed that scientific discoveries should belong to humanity as a whole. He never sought patents for his experiments, dedicating them all to the greater cause of science.

Hargrave's journey into the realm of invention began from a very early age. Born in England and migrating to Rushcutters Bay in Sydney he had always displayed an inventive spirit. Among his early creations were boat-shaped boots with hinged flaps, allowing him to walk on water—a whimsical yet ingenious testament to his imagination.

Hargrave's notebooks began brimming with visionary designs for aerial craft, some dating back to 1872. These concepts included ornithopters that imitated the graceful flapping of birds' wings.

In 1887, one of his ornithopters covered an impressive distance of 82 meters, a remarkable feat in the context of that time. By the end of 1892, Hargrave had developed up to 16 different flying machines, showcasing his unrelenting commitment to the dream of flight.

Throughout his life, Hargrave conducted a multitude of experiments and created countless models of his visionary ideas.

His curiosity knew no bounds.

He delved into the intricacies of aerofoils, developed a wave-propelled boat, designed a screw-driven engine, and crafted a working model aircraft with flapping wings. His ingenuity even led to the creation of a rotary engine so advanced that it wasn't until 1908 that it found its rightful place in aviation.

Hargrave's foresight extended well beyond the realm of invention. He predicted that man-made flight was not only inevitable but also that public demand for rapid travel would propel his dream into reality. However, he humbly cautioned *'The flying machine of the future will not be born fully fledged and capable of a flight for 1000 miles or so. Like everything else, it must evolve gradually.'*

Then, tragedy struck. In May 1915, Hargrave's son and fellow experimenter, Geoffrey Lewis Hargrave, perished at Gallipoli. The shock of this loss left Lawrence Hargrave seriously ill, and he passed away in a hospital in July 1915 from peritonitis.

A man of modesty, unassuming demeanour, and selflessness, Lawrence Hargrave left a profound legacy.

Upon his death at the age of 65, he was remembered as a *'gentleman of science... who probably did as much to bring about the accomplishment of dynamic flight than any other single individual'.*

Today, Stanwell Park remains a place where the winds still remind us of our past. It offers ideal conditions for hang gliding and paragliding, making it one of Australia's most famous venues for kite flying. Above the beach, a commemorative plaque stands as a testament to the historic event that unfolded in 1894, honouring Lawrence Hargrave, the unassuming visionary whose dreams took flight and forever changed the course of world aviation.

Harry Houdini

Pioneer of Power

The history of Australian aviation is a captivating tale of innovation, courage, and determination. While Lawrence Hargrave is often celebrated as the pioneer of flight in Australia, other inventors and aviators were not far behind. Among them, George A. Taylor and Colin Defries made significant contributions in the early 20th century.

However, it was Harry Houdini, the world-renowned

escapologist, who would achieve the first controlled flight in Australia to etch his name in the history of aviation.

George A. Taylor and Colin Defries: Early Attempts

On the 5th of December 1909, George A. Taylor achieved the distinction of becoming the first Australian to make a 'heavier-than-air' flight in a series of glider flights at Narrabeen. This marked a significant step forward.

Later that same month, Colin Defries attempted flight using an imported Wright brothers-designed Biplane. The stage was set at Victoria Park Raceway in Sydney on the 9th of December 1909. With much anticipation, Defries embarked on his journey, flying approximately 100 yards (about 91 meters) before losing control and crashing. Despite his first failed valiant attempt, he never gave up.

A second attempt on the 18th of December saw his mechanic, R.C. Banks, take to the skies. However, this too ended in a crash, dampening the spirits of those hoping for success.

While some considered the Colin Defries flight as the first in the Southern Hemisphere, others questioned its validity due to the limited time in the air and the subsequent crashes.

Within just twelve months of the Defries's attempt, other aviators armed with imported flying machines were determined to achieve Australia's first controlled powered flight. Several contenders made claims of success, but it was Ehrich Weiss, better known as escapologist Harry Houdini, who would ultimately succeed.

The Dawn of Australian Aviation. Harry Houdini's Historic Flight

On the 18th of March 1910, history was in the making. In the early morning light, a blanket of fog slowly lifted above a lonely paddock at Diggers Rest, located just 20 miles north of Melbourne. Two large tents had been erected side by side, housing two special planes shipped to Melbourne in preparation for his historic moment.

Two men had drawn up a list of 30 guests, one of whom was easily recognisable in the dawning mist as the world-renowned escapologist Harry Houdini. The other was pilot, Ralph C. Banks, and both were about to compete for the record of becoming the first to achieve a controlled powered aerial flight in Australia. Hargrave had already achieved flight but not in a controlled situation.

A wind had sprung up across the paddock as several hands helped Ralph Banks position his imported Wright Model A Flyer for take-off. He had won the right to make the first attempt.

Moments later the silence was broken as Banks wobbled down a slight incline before clearing the ground. He had barely covered 300 metres at a height of fewer than 5 metres when a sudden wind gust forced his plane into a dive, ending in several heavy thuds as his plane somersaulted across the paddock. Banks was thrown clear and escaped with only minor injuries but he was badly shaken and bitterly disappointed that his attempt had failed.

All eyes were now on Houdini who had brought along a journalist from the *Argus* to officially chronicle the attempt. The American showman, who had personally imported a French-built Voison bi-plane, clambered aboard.

It was now around 8:00 a.m. and while earlier attempts had been hampered by strong wind gusts, Houdini was confident that the conditions appeared to be easing.

The signal was given and the aircraft slowly picked up speed until it gradually lifted into the air under its power. It then flew a full circle of the paddock to the sound of the cheering guests before landing about a short time later.

Two further flights followed that same day lasting up to three and a quarter minutes in duration and reaching a height of 30 metres.

Houdini's Legacy.

Houdini, who was in Melbourne on a theatrical tour, had invited Australia's aviation pioneer Lawrence Hargrave to attend, but Hargrave was said to have dismissed the invitation out of hand in his '*typical acerbic manner*'. Hargrave was not an attention-seeking entrepreneur and his flight had been in the interests of science.

Houdini was a showman of the first ilk, and he had no false modesty about his achievements as a performer.

The great Harry Houdini held a special place in Melbourne having performed several death-defying stunts including a leap from Queens Bridge into the murky Yarra with his hands cuffed behind his back.

However, he regarded aviation strictly as a hobby. Unlike his career as an escapologist, he was never known to brag about his flying skills.

After his record flight at Diggers Rest, he wrote in his diary that he was '*...never in any fear and never in any danger; it is a wonderful thing*'.

The magic of flight, he would later say was in the '*...glorious thrill of the first adventure and not in a minor modification which is perpetual in any art*'.

The curious thing is that when he left Australia, he gave up flying altogether and sold the Voison aircraft. He never even drove a car again. He was, after all, only part of our history of flight in Australia.

Harry Houdini's achievement in aviation added another layer to his legacy. In Melbourne, where he had previously performed death-defying stunts, including a leap from Queens Bridge into the murky Yarra with his hands cuffed behind his back, he now held a new achievement as an aviator.

Despite his fame, Houdini remained modest about his flying accomplishments. He never used aviation as a means to further his showmanship but rather pursued it for the sheer joy of adventure and exploration.

In conclusion, the story of Harry Houdini's historic flight at Diggers Rest on March 18, 1910, remains a significant chapter in

Australian aviation history. It reflects the spirit of early aviators, the pursuit of the unknown, and the determination to achieve the seemingly impossible.

John Robertson Duigan

First Australian-built plane.

'To invent an aeroplane is nothing.
To build one is something. To fly is everything'
– Otto Lilienthal

On the 16th of July 1910, in the quiet Australian countryside near Kyneton, a momentous event took place. The tiny engine of an aircraft spluttered, coughed, and then fell silent. John Robertson Duigan and his brother Reg tinkered with the motor before giving the flimsy propeller another spin. This time, the sound was consistent and loud enough to disturb a flock of cockatoos from their home at 'Spring Plains', the Duigan family property.

Moments later, the contraption followed the startled birds into the sky, marking the first power-controlled flight of an all-Australian-designed and built aircraft.

This achievement occurred less than seven years after the Wright Brothers' historic flight in North Carolina and just four months after Harry Houdini's flight in a French-built aircraft not far from the Duigan property at Diggers Rest, Melbourne.

However, John Duigan never considered his first flight a controlled one. His brother Reg, who had assisted in building the aircraft, described it as more of a hop, a tentative tiptoe across the paddocks.

Undeterred by the challenges they faced, John and Reg continued their aviation experiments. Three months after their initial hop, on the evening of 7th October 1910, they attempted another flight. This time, their bi-plane covered 196 yards in a controlled flight path.

Whether it was a hop or a controlled flight, these early flights marked significant progress in Australian aviation history.

Despite the pioneering work of John Duigan, history began to portray him as a 'farm boy' who somehow cobbled together a flying machine from fencing wire and scraps. Reg Duigan vehemently disagreed with this depiction, asserting that John was a first-class engineer.

The legend may have overshadowed the facts, but the true facts record John Duigan as an Australian aviation legend.

John Robertson Duigan was born in Terang, Victoria's Western District, in 1882. He travelled to England, where he studied electrical and motor engineering at Finsbury College in

London. In 1908, he returned to Australia, and together with his younger brother Reg, they embarked on their aviation journey.

The frame of their first bi-plane was constructed from red pine and nearby mountain ash. John Duigan's ingenuity extended to designing the wheels of his aircraft, crafting his propeller shaft, and even creating the ball-bearing races.

He went as far as casting his water pump and making his radiator. An Engineering Company in Melbourne supplied a twenty-horsepower, four-cylinder engine, which Duigan meticulously redesigned for his specific purpose.

However, flying the aircraft was no small feat. The brothers needed outside help to transport it from their shed, down a steep hill, and across two creeks, for which they had to build special bridges to accommodate the size of their invention. At least three individuals were required to move the aircraft back and forth up the hill and back into the shed after each flight.

Following his successful 'hop' in July 1910, John Duigan applied to the Defence Department for details on a Commonwealth Prize of £5000 for the builder of a suitable military aircraft. Unfortunately, he missed the official entry deadline. Undeterred, he persisted, and on 7th October 1910, he achieved his first sustained flight of 178 meters at an altitude of about three meters, witnessed by a small group of spectators. For John, this marked his first successful flight under full control.

The Duigan brothers designed three different types of aircraft, with the third one built in the backyard shed of their parents' home in the Melbourne suburb of Ivanhoe.

Disappointingly, this two-seat tractor bi-plane crashed on its maiden flight at Keilor Plains in February 1913, and John Duigan suffered significant bruising. Plans to rebuild it were put on hold as the shadow of World War I loomed.

After John Duigan made further improvements by increasing engine power and refining wing balance he attempted to re-offer his aircraft design to the military. However, his offer was never accepted. The Defence Department did however request a demonstration a year later, but the outbreak of World War I diverted their focus.

Despite these setbacks, both John and Reginald Duigan continued to fly their aircraft locally, achieving distances of up to two kilometres and altitudes ranging from two to 18 meters. They completed a remarkable sixty flights before placing the aircraft in a hangar, as John enlisted with the Australian forces.

Although John's wartime service did not directly involve his aircraft expertise, he was recognised for gallantry in action with a Military Cross.

After the "war to end all wars", John Duigan returned to Australia. He established his motor engineering business, Old Bridge Motors, in Yarrawonga, which he ran until suffering a minor heart attack in 1928, prompting his return to Melbourne. He continued working until the outbreak of World War II, during which he offered his services and aviation expertise as an inspector of aircraft parts. He eventually retired to his home in Ringwood, where he succumbed to cancer on 11 June 1951.

In 1920, John Duigan generously donated his original biplane to the government, where it has since been preserved in the Victorian Museum collection.

To honour the contributions of both brothers to the aviation industry, Qantas named an Airbus A380 'Reginald Duigan'. Today a memorial stands at the site of that first flight on 16th July 1910.

It's a lasting testament to the history of flight and the path it paved for other Australian aviators to follow.

William 'Billy' Hart

Australia's First Official Aviator

William 'Billy' Hart, an Australian daredevil with a dream to fly, left an indelible mark on the early days of aviation in Australia. Born in Sydney in 1885, Hart's journey into the world of flight began as an apprentice to a dentist at the young age of 16.

However, his true passion lay in the early development of aviation, a field that was just beginning to take shape and he felt was out of his reach.

In September 1911, fate intervened as Hart crossed paths with a visiting British pilot touring Australia to demonstrate the latest Box kite from his company, the British and Colonial

Aeroplane Company. This chance meeting would alter the course of Hart's life.

Intrigued by the demonstration, he embarked on a bold new venture.

With determination and perhaps some financial wizardry, William Hart acquired the Bristol Box-Kite for a substantial sum, exceeding 1,300 pounds. Not only did he secure the aircraft, but he also received fundamental instructions from the company's mechanic on how to fly it.

However, as he was about to embark on his maiden flight with his newly acquired flying machine at Belmore Park, Penrith, tragedy struck. A sudden gust of wind wreaked havoc, destroying the fragile aircraft. Undeterred, Hart, with the assistance of friends and the company mechanic, embarked on the formidable task of rebuilding a new aircraft using salvaged parts from the wreckage.

On November 3, 1911, William 'Billy' Hart took a momentous step. He climbed aboard the aircraft he and his companions had painstakingly reconstructed for his first solo flight.

The following day, he achieved a historic milestone by flying from Penrith to Parramatta, with his younger brother Jack as a passenger. This marked the first cross-country flight in New South Wales. Such was the significance of this feat that, in 1963, a monument was erected in Parramatta Park to commemorate the historic spot where he landed.

Hart's achievements continued to soar. By November 16, 1911, he had successfully completed flying tests conducted by the Aerial League of Australia, who then presented him with Australian aviator's license no.1, dated December 5, 1911.

In doing so, he became the '*first airman to qualify as a pilot in Australia*'.

The following year, in 1912, Hart added yet another feather to his cap. He emerged victorious in Australia's first air race, covering the 32-kilometre distance from Botany to Parramatta in just 23 minutes. His skill and determination were evident as he continued to push the boundaries of aviation in his homeland.

Not content with his achievements, Hart aimed to share his passion and knowledge with others.

In the same year, 1912, he opened an Aviation School in Penrith, where he took on student aviators, nurturing the next generation of pilots.

Hart's pioneering spirit and relentless drive were further exemplified in August 1912 when he constructed a two-seat monoplane. Successfully testing it in Wagga Wagga showcased his engineering and aviation prowess.

However, tragedy was destined to strike again as the aircraft was wrecked in a serious accident at Richmond, leaving Hart with severe injuries that would ultimately ground him for life. It was a heartbreaking turn for the aviation pioneer who had once soared to such great heights.

In January 1916, Hart's life took another unexpected turn as he enlisted in the Australian Imperial Force as a lieutenant in the No.1 Squadron, Australian Flying Corps. Serving in Egypt and Britain as a non-flying instructor, he had to reconcile his passion for flying with the harsh reality of his physical limitations. Unfortunately, he was soon found to be medically unfit and returned to Australia, where he was honourably discharged on September 11.

After his military service, William 'Billy' Hart resumed his career as a dentist, undoubtedly a testament to his resilience and adaptability. Tragically, he succumbed to heart disease and passed away in Sydney on July 29, 1943.

His legacy, however, lives on in Australian aviation history. Hart's pioneering spirit, his contributions to early aviation, and his unwavering determination to conquer the skies continue to inspire generations of aviators in Australia and beyond. William 'Billy' Hart, Australia's first official aviator, will forever be remembered as a true pioneer who dared to dream and reached for the sky.

Legends of the Smiths

Australian Aviation Heroes

Sir Keith and Sir Ross Macpherson Smith: From England to Australia

In 1919 to further stimulate interest in aviation, the Commonwealth Government announced a prize of £10,000 for the first Australian to fly from England to Australia.

When WWI first broke out, all three Macpherson brothers from Adelaide immediately attempted to enlist.

Ross Macpherson Smith was accepted as a private in the 3rd

Light Horse Regiment and embarked for Egypt before landing on Gallipoli on 13 May 1915.

As a foot soldier on Gallipoli, he watched in awe as flimsy flying machines wove through the skies above observing enemy Turkish troops. This experience encouraged him to become an observer on reconnaissance aircraft but he soon began training as a pilot.

Keith Macpherson Smith's application to serve in the military was initially rejected on medical grounds. However, after he underwent medical treatment and later paid for his passage to England he enlisted into the Royal Flying Corps.

Within weeks of each other, they had both become aviators but in 1918 while Keith was posted to a newly formed bomber unit in France he was not to see active service and spent the rest of the war back in Britain with training establishments. His opportunity to become an Australian legend was just behind the next cloud.

Colin Macpherson Smith the third and youngest brother had his first application rejected due to his age. However, in June 1915 he was duly accepted and after a brief period was promoted to lieutenant in 1917. Tragically he died from war wounds inflicted at Passchendaele on the Western Front, a five-month campaign that cost the lives of half a million men.

* * *

The early career of Keith Macpherson Smith was far different from his brother Ross, yet both were to enter aviation within weeks of each other.

Before the outbreak of war, Ross had already launched an aviation career.

He had established a business of piloting joy flights in Australia while joining an aerial stunt team.

Yet, his true ascent into the aviation world only began to take serious shape when he formed a partnership with another pioneer aviator Charles Ulm establishing a mail transport service

Their journey began with a record-breaking Australia circuit with minimal navigational aids, earning them a £9,000 grant from the New South Wales Government. But Keith Macpherson Smith was not one to rest on his laurels; he aspired to push the boundaries even further.

It wasn't until July 1917 when Ross responded to a call for volunteers to join the Australian Flying Corps that Keith was inspired to take an active role in WW1.

He began by taking part in attacks, aerial photography missions, and bombing raids on Turkish forces. On one occasion he landed in the face of the enemy to rescue a downed comrade

He was also the pilot for T.E. Lawrence (Lawrence of Arabia) who described him as *"an Australian who delighted in taking additional risks"*.

During his extensive war service, Ross shot down 11 enemy planes and finished his service as Australia's most decorated WWI pilot. He was twice awarded the Military Cross, received the Distinguished Flying Cross three times, and the coveted Air Force Cross.

In 1919 the Australian Government offered a prize of ten thousand pounds for the first Australians to fly from Great Britain to Australia.

With news of the 'Great Race,' Ross Macpherson Smith once again joined up with his brother Keith and obtained a Vickers Vimy bomber, the same type of aircraft used by Alcock and Brown to cross the Atlantic earlier that year.

With Keith as assistant pilot, a navigator and accompanied by two mechanics, Wally Shiers and Jim Bennett, they left Hounslow, England, on 12 November 1919.

Their flight was nothing short of extraordinary, breaking the record and establishing their reputations worldwide.

They arrived in Darwin, Australia, in just 28 days; two days ahead of their ambitious target. Their total flying time amounted to a mere 136 hours.

In addition to claiming the £10,000 prize, they became the first Australians to fly from England to Australia, achieving legendary status in the process. The recognition of their achievement extended to knighthoods for both brothers.

However, the story of the Smith brothers doesn't end with this incredible feat. Their passion for aviation led them to their next adventure: a journey around the world in a Vickers Viking amphibian aircraft.

Both brothers travelled to England to prepare for the trip but on 13 April 1922, while Ross and his long-serving crew member Jim Bennett were test-flying the aircraft at Weybridge near London, it spun out of control and into the ground from 1000 feet (305 m), killing them both instantly. Keith, who arrived late for the test flight witnessed the accident.

The investigating committee concluded that the crash had been the result of a pilot error. It was revealed that Ross had not flown for many months before this test flight and had never flown that type of aircraft before.

After their untimely deaths, the bodies of Sir Ross Macpherson Smith and Lieutenant Jim Bennett were returned to Australia. They received a state funeral, and Sir Ross was laid to rest in Adelaide on June 15.

Despite the tragic loss of his brother, Keith maintained his interest in aviation.

One promising venture he strongly supported in the early 1920s was to employ Vickers-built airships on Imperial air routes.

A British airship had successfully crossed the Atlantic in July 1919, but further development of the project failed to materialize.

Sir Keith Smith continued to maintain a connection with Vickers serving as the Australian agent for the aircraft. He was to become vice president of British Commonwealth Pacific Airlines and a director of Qantas and Tasman Airways. By the end of his career, Keith was in control of the many Australian-based Vickers companies.

Sir Keith Smith died of cancer in Sydney on 19 December 1955 leaving an estate valued at £33,723 and an enduring legacy. His Vickers Vimy, flown on that historic flight, remains on display at Adelaide airport.

Sir Charles Kingsford Smith

The Boldest of All

Among the many pioneers who etched their names into our history books one name stands out with boldness and daring – Sir Charles Kingsford Smith, affectionately known as 'Smithy'.

If Sir Lawrence Hargrave is remembered as the 'first to fly' then it's fitting to consider Smithy as the man who carried the torch of aviation into a new century.

Of all the pioneer aviators in Australia's history Sir Charles

Kingsford Smith or 'Smithy' as he is affectionately remembered, is regarded as the boldest flyer of them all.

The feats of Charles Kingsford Smith have been likened to the modern-day equivalent of Neil Armstrong's moon landing.

Today the legacies of Hargrave and Smithy are enshrined on the Australian $20 banknotes, a testament to their pivotal roles in shaping the nation's aviation heritage.

* * *

Charles Kingsford-Smith was born in Brisbane on 9 February 1897. When WW1 broke out "Smithy" answered the call and enlisted in the Australian Army. He played a crucial role as a motorcycle dispatch rider during the challenging Gallipoli campaign. Later, he transitioned to the Royal Flying Corps, successfully earning his pilot's wings. While serving in France, having already shot down four German aircraft, a fateful turn of events in 1917 saw his aircraft shot down.

His extraordinary courage led to the award of the Military Cross. Due to injuries, which included the amputation of two toes, he was granted permission to return to Australia to recover, where he stayed with his parents.

Upon making a complete recovery, he returned to England and reengaged in the war effort, serving as a flying instructor.

Post-war, Smithy chose to remain in Britain and embarked on an entrepreneurial journey alongside fellow Australian Cyril Maddocks. Together, they procured surplus aircraft, offering thrilling joyrides to the public. Smithy later ventured to the United States, where he worked in varying capacities as a pilot, including a brief period as a stunt flyer in California, before joining the fledgling aviation industry back in Australia.

In 1927 he partnered up with a fellow flying enthusiast Charles Ulm to become the first airmen to fly around Australia.

The following year, with the support of wealthy businessmen and government grant money – Kingsford-Smith, Ulm together with two Americans, Harry Lyon and James Warner as navigator and wireless operator, became the first airmen to cross the Pacific.

The flight was described as particularly dangerous with US betting odds of eleven to one of even reaching Hawaii. Ten fliers had perished the previous year attempting just the first leg of the Pacific crossing.

The Southern Cross had been purchased second-hand from Australian polar explorer George Hubert Wilkins after it had been salvaged from a crash in Alaska in 1926. Fitted with three new engines, courtesy of Melbourne businessman Sidney Myer the noise from the engines was described as so deafening that the crew were forced to communicate with each other by messages scribbled on scraps of paper. By the time they reached Fiji, the crew were stone deaf.

Despite storms, rain squalls, instrument failures and a heart-stopping eight-minute engine splutter mid-ocean, the Southern Cross and its crew finally completed the 12,000-kilometre journey after a total of 83 hours flying time.

Their arrival in Sydney on 10 June 1928 was met by a huge crowd estimated at 300,000. Police were forced to cordon off the plane in a bid to prevent the excited crowds from rushing into the spinning propellors. By now they were being treated like film stars.

The following August, Kingsford Smith and his crew flew the Southern Cross from Point Cook in Victoria to Perth, then shortly afterwards, from Sydney to Christchurch, becoming the first airmen to cross the Tasman Sea.

But on March 31, 1929, during a planned return flight from

Sydney to England, the 'Southern Cross,' with Smithy at the helm, encountered a crisis.

The aircraft was forced to make an emergency landing on a mudflat in the Kimberly region of north Western Australia, triggering a massive search operation. Tragically, two of Smithy's friends, who were on their way to assist in the search, crashed in the Tanami Desert in Central Australia and lost their lives. Much of the blame for the incident was unjustly placed on Kingsford Smith.

Although he was eventually exonerated by an inquiry, the incident left a stain on his reputation in Australia. Despite this setback, Smithy's passion for adventure endured. In June of that same year, he resumed his flight, ultimately reaching England in a record-breaking 12 days and 13 hours of flying.

Adventure flying was Kingsford Smith's true passion, and in 1930, he undertook another daring journey. This time, he flew the 'Southern Cross' across the Atlantic from England to Oakland USA, becoming the first person to circumnavigate the globe by air.

His insatiable thirst for records and challenges continued, as he embarked on a long-distance flight from England to Australia in an Avro Avian biplane named the "Southern Cross Junior." This flight shattered Bert Hinkler's solo record by an impressive five-and-a-half days.

Records continued to fall with Smithy's relentless determination. However, his pursuit of aviation milestones took a toll on his burgeoning airline business.

On March 21, 1931, the 'Southern Cloud', flying from Sydney to Melbourne with a pilot, co-pilot, and six passengers, vanished in severe storms over the Snowy Mountains. Tragically, there were no survivors, and the wreckage remained undiscovered until 1958.

In 1932, he was knighted in recognition of his outstanding contributions to aviation services.

However, the loss of his brother, combined with the deepening grip of the Great Depression, had dealt a body blow to Smithy's health. He found himself almost back to where he started, offering joy flights for ten shillings a trip. Then, toward the end of 1933, his prospects began to brighten.

In 1934, Smithy purchased a Lockheed Altair and named it the 'Lady Southern Cross' to compete in the MacRobertson Air Race. Unfortunately, he couldn't make it to England in time for the start of the race.

Undeterred, he embarked on another historic journey, making the first eastward crossing of the Pacific Ocean by an aircraft, flying the 'Lady Southern Cross' from Australia to the United States.

At the age of 38, while the toll from his arduous career and daring exploits were further impacting his health, his passion for flight still burned.

He left the 'Lady Southern Cross' in the US in the hope of selling it.

However, his commitment to demonstrating the future of global air transport remained unwavering. Unable to find a purchaser, he arranged for the unsold 'Lady Southern Cross' to be shipped to England from where he took off on November 6, 1935, aiming for one more record-breaking flight to Australia.

During the final leg of this last epic flight, the 'Lady Southern Cross' with Sir Charles Edward Kingsford Smith, M.C., A.F.C., and co-pilot John Thompson Pethybridge were lost.

Their disappearance occurred somewhere off the coast of Burma while flying at night toward Singapore. A witness reported seeing a plane battling a storm 150 miles from shore, flying 200 feet over the sea with fire erupting from its exhaust.

Eighteen months later, the wreckage was discovered and

confirmed to be from the 'Lady Southern Cross'. However, no bodies were ever found.

Sir Charles Kingsford-Smith has been called the world's greatest aviator. His record-breaking flights and almost superhuman flying skills are legendary.

For his contribution to civil aviation, he was knighted in 1932.

While Sir Charles Kingsford Smith is rightfully remembered as a legend in Australian history, his fame tended to overshadow the daring exploits of another aviation family bearing the name of Smith.

Herbert John Louis Hinkler – Bert Hinkler

Bert Hinkler: The Australian Lone Eagle

Bert Hinkler was not only a prolific inventor but as a solo aviator he became affectionately known as the "**Australian Lone Eagle**".

He was designing and building aircraft well before becoming the first person to fly solo from England to Australia, and the first person to fly solo across the Southern Atlantic Ocean.

In 1928 few names in aviation shone as brightly as that of Bert Hinkler.

Bert's extraordinary journey into the world of aviation began in the Queensland town of Bundaberg on December 8, 1892. From humble beginnings, he would go on to achieve ground-breaking feats, designing and building early aircraft, becoming the first person to fly solo from England to Australia, and securing his place as the first person to fly solo across the formidable Southern Atlantic Ocean.

Hinkler's fascination with flight was kindled at a remarkably young age.

According to his mother, at just five years old, he pointed to a graceful ibis soaring through the skies and expressed a heartfelt desire to fly like that bird. It was a dream shared by many young boys, but Bert's determination set him apart. Undeterred by the laws of physics, he even attempted to take flight himself by strapping homemade wings to his back. While these early efforts were unsuccessful, they revealed a passionate curiosity that would define his life.

By the tender age of 11, Hinkler's passion for aviation took flight when he learned of the Wright Brothers' pioneering flight in America.

This revelation ignited a further passion, leading him to collect every available article about aviation from newspapers and magazines.

His fascination with flight drove him to leave formal education at the age of 14. Instead, he spent his time tinkering with gliders and other contraptions, all in pursuit of his dream to conquer the skies.

By the time he turned 19, Bert Hinkler had soared to new heights—both literally and metaphorically. He had successfully piloted his homemade glider to a remarkable altitude of 33 feet (10 meters). This early achievement further fuelled his determination to master the art of flying.

In 1911, he enrolled in an aviation correspondence course, while continuing his experiments with gliders. His journey was slowly taking shape transitioning from a childhood dream to a fully-fledged reality.

Bert's destiny took an intriguing turn when American pilot Arthur Burr "Wizard" Stone arrived in Bundaberg in 1912 with his travelling flying show. It was a momentous occasion for the young Hinkler, as it marked the first time he had ever witnessed an aeroplane in action. In a remarkable twist of fate, Hinkler identified an instability issue with Stone's Bleriot aircraft and provided a solution.

Impressed by Hinkler's knowledge and resourcefulness, Stone offered him a job as a mechanic. This chance encounter provided Bert with valuable hands-on experience.

However, fate had other plans in store. When Stone's tour abruptly ended with a crash in New Zealand, Hinkler found himself at a crossroads. His unwavering determination to pursue aviation led him to accept a job at an aeroplane factory in Penrith. This experience honed his skills further and deepened his understanding of aircraft construction and mechanics.

Believing that Europe had become the epicentre of aviation, Bert Hinkler made a momentous decision in 1914 and embarked on a journey to England. His goal was clear: to work in the heart of the aviation industry and realise his dream of becoming a pilot. He found a job at Sopwith Aviation, a prestigious aircraft manufacturer, which not only provided him with invaluable experience but also supported him in obtaining his pilot's license.

However, the world was on the brink of war, and in August of that year, World War I erupted.

Undeterred by the chaos unfolding around him, Hinkler decided to channel his passion for aviation into serving his country.

He joined the Royal Navy Air Service (RNAS), in another bid to earn his wings.

During his time in the RNAS, Hinkler continued to showcase his innovative spirit by developing a dual-control system that allowed a gunner to take control of the plane if the pilot was injured or killed. This invention demonstrated his commitment to improving the safety of aviation, a theme that would reappear in his later work.

Hinkler's contributions extended beyond the dual-control system. He also made several improvements to aircraft, including designing machine guns that could eject shells, thus preventing them from hitting the pilot—a perilous risk that had plagued many aviators.

Bert Hinkler's journey then took another turn in 1917 when he finally earned his pilot's license.

After the war, he was posted to Italy with the newly formed Royal Air Force (RAF). It was during this time that he continued to hone his skills and deepen his knowledge of aviation.

In 1919, Australian Prime Minister Billy Hughes announced the Great Air Race, offering a substantial £10,000 prize for the first successful flight from England to Australia. Hinkler, ever the aviator, eagerly applied as a solo pilot which would test not only his flying skills but also his courage and endurance. However, organisers of the race, which was eventually won by brothers Ross and Keith Smith, deemed Hinkler's solo flight too risky and barred him from competing.

Yet, Hinkler remained undeterred. If anything, it just inspired his determination to prove himself as a solo aviator.

In 1920, he made his first solo attempt in an Avro Baby. Unfortunately, this daring endeavour had to be abandoned in the Middle East due to unforeseen challenges.

Still, Bert Hinkler's spirit was unbreakable. He eventually acquired an Avro Avian G-EBOV, a small biplane that would become his trusted companion on several historic journeys.

On the cold morning of February 7, 1928, he took off from Croydon, England, with minimal fanfare. Accompanied only by his wife and two witnesses.

Hinkler was focused on one thing: breaking the record set by the Smith brothers in 1919 and becoming the first person to make the journey solo.

His flight path took him across Asia and the Timor Sea, using a London Times Atlas as his primary navigational chart. The sheer audacity of flying solo across vast distances with such rudimentary navigation tools was a testament to his determination. In under 16 days, he completed the journey, arriving in Darwin on February 22, 1928, after flying for a total of 128 hours.

Between his arrival in Darwin and his triumphant landing in Sydney on March 11, 1928, Hinkler realised the importance of publicity. He began actively informing the press of his progress, and his fame skyrocketed. The public affectionately dubbed him 'Hustling Hinkler', a nickname that would be immortalised in song. His arrival in Sydney was met with an outpouring of support, with an estimated 80,000 people gathering to welcome their hero, singing, *"Hinkler, Hinkler little star, sixteen days and here you are."*

Bert Hinkler's remarkable triumph in 1928 was lauded around the globe. Commercial clothing manufacturers swiftly capitalised on his popularity, introducing the 'Hinkler' hat as a fashionable accessory.

The Australian government also recognised his extraordinary achievement by awarding him a £2000 reward, which he used to finance more dreams, including building his own aircraft.

Hinkler had become a national hero, joining an exclusive club of pioneer aviators.

In 1931, Bert Hinkler embarked on perhaps his most remarkable feat. He flew from Canada to New York, completing a non-stop flight to Jamaica, and then crossed the treacherous South Atlantic to Africa. This remarkable solo journey marked the first-ever flight across the South Atlantic by a lone pilot. It was only the second solo Atlantic crossing since Charles Lindbergh's historic flight in 1927.

However, as with many pioneers of the skies, fate always seemed to loom on the horizon. In 1933, while attempting to break the Britain to Australia record once more, Hinkler's plane tragically crashed into the Pratomagno Ranges in Italy.

At the age of 40, Bert Hinkler had reached the end of his remarkable journey.

He was laid to rest with full military honours in Italy, on the orders of his greatest admirer in that country, Benito Mussolini.

The legacy of Bert Hinkler endures through the pages of aviation history.

His pioneering solo flights, innovative technology, and fearless pursuit of adventure continue to recognise the courage of those pioneers who also helped shape Australia's aviation industry.

Today, his legacy lives on through institutions like the Australian Dictionary of Biography, the State Library of Victoria, and the Hinkler Hall of Aviation. Bert Hinkler's remarkable life story is a reminder that dreams when fuelled by unwavering determination, can leave an indelible mark on the world.

Charles Ulm

Early Aviator 1928

Charles Ulm's story is another of those remarkable pioneer adventures.

Despite being overshadowed by fellow aviator Sir Charles Kingsford Smith he joined him in partnership, sharing similar visions and determination. Like Smithy, his efforts to establish trans-Pacific air services and his contributions to record-breaking flights also left a lasting legacy.

A Visionary Pioneer of Australian Aviation

The name Charles Ulm may not resonate as loudly as some of his contemporaries, but his contributions to the world of flight are no less profound. Born in Middle Park, Melbourne, in 1898, Ulm's journey into aviation was anything but conventional, marked by a series of audacious decisions and a relentless pursuit of his passion for flying.

A False Start and Gallipoli

Ulm's early life was marked by a thirst for adventure and in 1915, at the tender age of 16, he enlisted in the Australian Imperial Force (A.I.F.), using a false name and age to circumvent the enlistment regulations.

His towering stature, at almost six feet, lent credibility to his alleged age of 20. Little did he know that this daring move would set the stage for his remarkable career.

Ulm's journey took an unexpected turn when he found himself among the first troops to land at Gallipoli on April 25, 1915. Amid the battle, he displayed remarkable courage but was soon wounded in action. This marked the end of his first stint in the A.I.F.

Still a minor he was discharged at the request of his parents.

The Western Front

In 1917 Charles Ulm reenlisted, this time under his true name and age. He returned to the Western Front, where the horrors of war once again tested his mettle. Tragically, he was once more badly wounded, leading to his evacuation to Britain and eventual demobilisation in March 1919.

While awaiting repatriation in England, Charles Ulm experienced a life-altering moment.

He found himself in control of an unauthorised aircraft, and while lacking any formal training he miraculously managed to land it without catastrophe.

This serendipitous moment ignited a hidden passion for flying and although he didn't earn his pilot's license until many years later, the episode served as the catalyst for his belief in the future of commercial aviation, particularly in isolated areas of Australia.

The Birth of a Dream: Trans-Pacific Flight

In 1921, Ulm embarked on his first attempt to secure funding for a trans-Pacific flight route. His vision was ahead of its time bridging the vast expanses of the Pacific Ocean. However, it wasn't until 1927 that his dream began to materialise when he crossed paths with a like-minded aviator, Sir Charles Kingsford Smith.

A Partnership That Soared

Ulm and Kingsford Smith joined forces in 1927, forming a pivotal moment in Australian aviation history. They began by circumnavigating Australia in record time that year.

However, their most iconic achievement came in 1928 when they co-piloted the historic flight of the Southern Cross from California to Brisbane, marking the first successful crossing of the Pacific Ocean by air.

Their accomplishments continued to mount as they made the first trans-Tasman flight from Sydney to Christchurch later in 1928. Ulm's unwavering determination and Kingsford Smith's flying skills made them an indomitable duo in the world of aviation. Together, they continued to push the boundaries.

Record-Breaking Flights

Charles Ulm's thirst for records and adventure remained unquenchable. In addition to their feats in the Southern Cross, he and Kingsford Smith set a speed record from England to Australia in their aircraft 'Faith of Australia', completing the journey in a mere 6 days, 17 hours, and 56 minutes. These endeavours solidified their status as pioneers of long-distance air travel.

Troubles on the Horizon

By November 1931, the partnership between Ulm and Kingsford Smith began to show signs of strain.

Dark clouds loomed over their joint company as challenges emerged on multiple fronts. Tragedy struck when an ANA aircraft named the Southern Cloud disappeared during a flight from Sydney to Melbourne on March 21, 1931. All on board perished, and the wreckage, along with the bodies of the pilot, co-pilot, and six passengers, remained undiscovered for a staggering 27 years.

The Unwavering Vision

Despite financial setbacks and the demise of ANA, Charles Ulm remained resolute. His gaze remained fixed on the vast expanse of the Pacific Ocean, believing that air travel was the future.

In September 1934, he established a new company, Great Pacific Airways, with himself as the managing director.

Facing financial constraints, the company acquired an

Airspeed Envoy aircraft and on December 3, 1934, 'Stella Australis' took off from San Francisco for Honolulu, with Ulm as the chief pilot.

Accompanied by a relatively inexperienced co-pilot and navigator, the journey was fraught with challenges. The crew sent Morse code messages indicating they were lost and running out of fuel, and it is believed they eventually ditched into the sea.

Despite exhaustive searches, no trace of 'Stella Australis' was ever found.

A Lasting Legacy

In the years since his death, his achievements have often been overshadowed by his partner, Sir Charles Kingsford Smith. However, history reveals that Charles Ulm was not just a pioneer aviator; he was a pioneer of Australian aviation itself paving the way for the growth of modern global air travel.

As one biography aptly recorded, Charles Ulm's legacy endures and deserves his rightful place alongside all his peers who lost their lives attempting to push the boundaries of human flight.

Sir Patrick Gordon Taylor

The Man Who Saved Smithy

Of all our early aviators who pushed the boundaries of human flight, none quite epitomised aerial heroics and bravery more than Patrick Gordon Taylor – immortalised as 'The Man Who Saved Smithy'.

Born in Sydney on November 7, 1899, Gordon Taylor developed an extraordinary spirit of adventure and an early fascination with flight. From the tender age of four, in an era when aviation itself was in its infancy, he nurtured dreams of soaring among the clouds—dreams that faced a stern reality check when, at the age of 18, he contracted tuberculosis, gravely threatening his health, if not his life.

Miraculously, he not only recovered from his illness but emerged with an unwavering determination to conquer the skies.

In August 1916, at the age of 20, he embarked on a journey to England, aiming to join the Royal Flying Corps. He would soon find himself in the heart of World War I, flying with the No. 66 Squadron.

In 1917, his exceptional skills and bravery earned him the prestigious Military Cross, and he was promoted to the rank of Captain. Taylor continued his service, flying with the 94th and 88th Squadrons. However, the horrors of war hit close to home when his brother Kenneth met a tragic end on assignment in France. This loss left Patrick deeply affected, and he wrote about how he 'deplored the killing and all the other evils of war'.

After the war in 1919, Taylor returned to his homeland, Australia, where he pursued his passion for flight as a private pilot. After completing an engineering course, he became proficient in aerial navigation.

He even operated a Gipsy Moth seaplane from Sydney Harbour between 1928 and 1932.

Additionally, he served as a captain with Australian National Airlines Ltd from 1930 to 1931. By this time, Australian aviation was experiencing a renaissance, and Taylor played a significant role in its development.

In 1925 Taylor's commitment to aviation led him to join the Royal Australian Air Force (RAAF) where he quickly demonstrated remarkable skill and bravery as a pilot. However, despite his achievements and contributions to aviation, like many of his peers, Taylor's name was often overshadowed by the more famous figures of Sir Charles Kingsford Smith and Charles Ulm.

Little did he know that he would soon have the opportunity to fly alongside both of his heroes.

By 1933, Taylor had become the second pilot and navigator for the famous Fokker Southern Cross, under the command of Sir Charles Kingsford Smith. Together, they embarked on historic flights in 1933 and 1934, covering routes between Australia, New Zealand, and England.

In the same year, Taylor also served as a navigator aboard Charles Ulm's aircraft, *Faith in Australia*, for two flights, connecting Australia to England and then returning.

Despite missing the 1934 Victorian Centenary Air Race, Taylor and Kingsford Smith achieved an even more monumental feat: completing the first-ever flight from Australia to the United States. This ambitious journey took them via Suva and Hawaii in the *Lady Southern Cross*, marking a milestone in aviation history.

On May 15, 1935, Taylor once again found himself flying alongside Sir Charles Kingsford Smith, this time as the navigator in the *Southern Cross* for the King George Jubilee airmail flight between Australia and New Zealand.

After six hours in the air, disaster struck when part of the centre engine's exhaust manifold broke off, severely damaging the starboard propeller. With the aircraft heavily laden, 'Smithy' made the difficult decision to shut down the malfunctioning starboard engine, jettisoning his cargo and applying full power to the remaining two engines before turning back to Australia.

The oil pressure on the port engine began to fall alarmingly, and the flight appeared doomed.

It was at this critical moment that Patrick Gordon Taylor's heroism shone through.

He climbed out of the fuselage, braving the strong slipstream, wove his way through the engine strut and began collecting oil from the disabled starboard engine into a thermos flask.

With the collected oil he then transferred it into the port

engine. With the assistance of the wireless operator, John Stannage, he repeated this procedure six times, all while the fate of the flight hung in the balance. Miraculously, their efforts paid off, and the aircraft landed safely at Mascot after a harrowing nine-hour ordeal.

For his resourcefulness and courage, Taylor was awarded the Empire Gallantry Medal in 1937, and in May 1941, he received the George Cross Medal, recognising his extraordinary bravery and heroism.

In 1943, Taylor's dedication to aviation led him to become a commissioned flying officer in the RAAF, eventually transferring to the Royal Air Force in 1944. Throughout the Second World War, Captain Taylor served as a pilot with the Air Transport Auxiliary, having now served in two World Wars.

His remarkable achievements continued even beyond the war. In 1951, he was honoured with the Oswald Watt gold medal for his Australia-South America flight, and he received the Johnson Memorial trophy from the Guild of Air Pilots and Air Navigators in London.

Finally, in 1954, he was knighted and became known as Sir Gordon.

Sir Patrick Gordon Taylor's adventurous life came to an end on December 15, 1966, when he passed away in Queen's Hospital, Honolulu. His ashes were scattered over Lion Island located at the mouth of the Hawkesbury River in Broken Bay on the Central Coast of New South Wales.

This was where the dreams that fuelled his extraordinary journey were first conceived.

Sir George Hubert Wilkins

First polar explorer in 1928

Of all the exploits and acts of sheer bravery by Australia's pioneer aviators, the life of George Hubert Wilkins would read like pure fiction had they not been so meticulously recorded as fact.

Born in South Australia in 1888, the official biography of George Wilkins traces a career which spans journalism, war hero, polar explorer, submariner, naturalist, climatologist and above all, aviator. And if truth be known he may also have been a spy adding an aura of intrigue to his already extraordinary life.

One of the most remarkable chapters of this epic unfolded during an incredible journey around the world in 1929 aboard the

German-built Airship LZ 127 Graf Zeppelin. This luxury airship, a marvel of its time, invited a diverse and international group of media professionals and privileged guests to partake in a ground-breaking three-week voyage around the globe.

Among the media contingent was the recently knighted Australian celebrity explorer and cinematographer, Sir George Hubert Wilkins, who was reporting for the American media empire of William Randolph Hearst. Little did Wilkins know that this journey would connect him with a Japanese observer who would play a significant role in his life years later.

Wilkins had already gained worldwide recognition as a pioneer aviator by conducting the first powered flight across the Arctic Circle, paving the way for further exploration of the North Pole.

When World War I erupted, Wilkins assumed the role of an official photographer, tasked with providing "an accurate and complete record of ground fighting and other activities for the AIF".

During this campaign, he displayed exceptional bravery and leadership, earning the Military Cross when forced to assume command of American troops who had lost their officers in the Battle of the Hindenburg Line.

Notably, he became the only official Australian photographer from any war to receive a combat medal, with Australian General Monash describing him as "the bravest man I have ever seen".

When the war ended, Wilkins continued to pursue aviation, entering the England-Australia Air Race in 1919. However, after surviving the crash of his aircraft he then joined Shackleton's 1921 expedition to the Antarctic.

During a series of flights over the Arctic Circle in 1926–27, Wilkins faced numerous life-threatening situations. In one harrowing incident, his plane ran out of fuel, forcing him to make an emergency landing on the sea ice. He and his crew were stranded for 13 days before being rescued, showcasing his remarkable survival skills.

The following year, Wilkins accomplished another astonishing feat by flying across the Arctic Sea, covering 2,500 miles of mainly uncharted territory. This achievement earned him a knighthood, and from that point forward, he chose to be known as Sir Hubert, rather than Sir George.

Never one to rest on his laurels, in 1931, Wilkins became the first person to venture beneath the Arctic ice.

As an Australian, he faced challenges forcing him to acquire a submarine in the USA. He overcame this obstacle by leasing a surplus WWI submarine for just one dollar. He renamed it 'Nautilus'. in homage to Jules Verne's fictional submarine from "Twenty Thousand Leagues Under the Sea".

Wilkins's polar expedition with the Nautilus faced its first loss before even leaving New York Harbor, as a crew member tragically drowned.

Soon after commencing the expedition, both the starboard and port engines of the submarine broke down. Stranded without propulsion, Wilkins sent out an SOS and was eventually towed to England for repairs.

On June 28, the Nautilus was once again operational, and it arrived in Norway to begin its journey beneath the ice caps. However, another setback awaited. The submarine was missing its diving planes, crucial for controlling it while submerged. Undeterred, Wilkins made the most of the situation, focusing on taking core samples of ice while studying salinity and gravity.

As he persevered with this challenge he received a wireless plea from his backer, William Randolph Hearst. Hearst urgently implored Wilkins to return to safety and postpone further adventures for a more favourable time and with a better-equipped vessel.

Heading back to Britain, Wilkins encountered a fierce storm, leading to severe damage to the Nautilus forcing him to make the difficult decision to deliberately sink the submarine.

Despite his obvious failure, he was able to prove that submarines were capable of operating beneath the polar ice cap. The scientific results of his limited efforts paved the way for future successful missions.

Less than a decade later, as the world teetered on the brink of World War II, the media baron Hearst once again enlisted Wilkins, this time for a thinly disguised espionage mission to observe the Japanese.

Wilkins had already travelled extensively in China, Japan and Asia gathering crucial information that suggested war in the Pacific was imminent.

It was during a fateful encounter in the famous bar at the Raffles Hotel in Singapore that Wilkins reconnected with his former Japanese friend from the Graf Zeppelin journey. This friend had ascended to the position of Japanese Consul General. As their conversation unfolded over drinks, the Japanese diplomat made a startling revelation: "I know what you're doing here. You're not on any economic survey; you're working for the US Government".

The inebriated diplomat continued, "Now I'll tell you something you don't know. Eighteen months from now, we will attack the American fleet in Pearl Harbor."

Wilkins was staggered by this disclosure. He inquired about the source of this information and the reason for sharing it. In response, the inebriated diplomat burst into laughter, challenging Wilkins to report the warning. He concluded with a sardonic remark: "They'll never believe a man who was crazy enough to try to take a submarine to the North Pole!"

The preposterous-sounding warning, delivered under the influence of alcohol, could not be ignored. Wilkins did report the information, but it seems that, tragically, it was either disregarded or dismissed as implausible.

On December 7, 1941, just before 8 a.m., the Japanese launched a devastating attack on Pearl Harbor.

Following the attack, Sir George Hubert Wilkins remained in the United States, dedicating his efforts to the war effort.

Importantly, he never renounced his Australian citizenship, of which he was intensely proud.

In 1958, at the age of 70, Sir George Wilkins passed away suddenly from a heart attack in Massachusetts.

In a fitting tribute to his adventurous spirit and contributions to exploration, his ashes were taken on the nuclear submarine Skate and scattered at the North Pole.

Today, numerous geographical features bear his name, including Wilkins Island, Wilkins Sound, Wilkins Ice Shelf, and Wilkins Aerodrome near Casey station. These enduring landmarks serve as a permanent testament to one of Australia's greatest aviators and explorers.

In summary, George Hubert Wilkins lived a life that defied imagination, remembered for his remarkable achievements in aviation, exploration, and wartime service. His encounters with influential figures like William Randolph Hearst and the Japanese Consul General during the turbulent times of World War II add layers of complexity to his story. Wilkins' warning about the Pearl Harbor attack, despite its unlikely delivery, stands as a haunting episode in his incredible life journey, reminding us of the unpredictable twists of history.

Sir Lawrence Wackett

'The Father of Australian Aviation'

Sir Lawrence Wackett is another of those influential figures in the history of Australian Aviation.

His contributions spanned a wide spectrum of activities from fighter pilot, aircraft and engine designer, entrepreneur, and leader of Australia's burgeoning Airforce.

His journey began during the turbulent era of World War I when he displayed not only courage as a fighter pilot but also demonstrated an innate talent for innovation and leadership.

Wackett was among a select group of twelve pilots dispatched to Egypt with the Australian Flying Corps, as part of the newly formed Australian Imperial Force (AIF). This elite group was formed just a day before Wackett celebrated his 20th birthday. It was here that he first demonstrated his resourcefulness and mechanical ingenuity. Remarkably, he crafted an aircraft machine gun using salvaged parts from a Singer sewing machine he had acquired on the streets of Port Said.

During his time in Egypt, Wackett's inventive mind was further showcased as he designed a mounting system that securely attached a Lewis Machine Gun to the upper wing of his aircraft.

This ingenious yet simple modification eliminated the risk of the gunner accidentally hitting his pilot significantly enhancing the effectiveness of aerial combat. Wackett's modified BE2c aircraft soon proved to be a formidable asset in numerous missions.

One particular incident highlighted his courage and leadership. During a reconnaissance mission, he fought off an enemy aircraft attack. Moreover, he embarked on a daring 7-hour bombing sortie to Beersheba, where his formation encountered two highly advanced German aircraft. Once again, he and his fellow aviators managed to fend off the attackers.

For his exceptional contributions and valour in battle, Wackett was rightfully awarded the Distinguished Flying Cross, an acknowledgement of his outstanding service during the war.

Following his service in Egypt, Wackett was transferred to No. 3 Squadron AFC in France, where he reached the rank of captain. It was during this period that he devised a compact parachute for dropping supplies to isolated troops on the ground. This inventive creation was later adapted into a bomb rack, proving highly effective against enemy ground forces. Another example of innovation that further solidified Wackett's reputation.

In a daring act of reconnaissance, Wackett penetrated six

miles behind enemy lines, capturing valuable aerial photographs of German forces below. Just two days later, he carried out an ammunition resupply mission to isolated allied troops, once again utilising the modified equipment he had designed. These audacious actions earned him a second Distinguished Flying Cross, underscoring his exceptional contributions during World War I.

On January 6, 1919, two months after the end of World War I, Wackett received a promotion to the rank of major and was appointed as the Commanding Officer of No. 7 Squadron, based in England. His leadership in this role further exemplified his dedication to aviation. However, he then made the difficult decision to return to his native Australia in March of that year, marking the end of his military service.

In the post-war era, Wackett emerged as one of the pivotal figures in the establishment of the Royal Australian Air Force in 1921.

His vision extended beyond the military realm; he aspired to create Australia's first aircraft industry.

To achieve this, he continued to innovate, introducing aircraft like the Warbler, a wing monoplane powered by an engine of his creation.

In 1924, he introduced the Widgeon I and II flying boats, followed by the Warragal I and II land planes. These aircraft significantly contributed to the growth of Australia's aviation industry.

The outbreak of World War II, saw Wackett once again display his unique ability of creation. He designed the Wirraway and presided over a company that employed thousands of people, delivering hundreds of aircraft, engines, propellers, and aircraft components to support the war effort. His contributions to the war were invaluable, and his dedication was unwavering.

Amid these professional achievements, Wackett endured a deeply personal tragedy when his son, Squadron Leader Wilbur Lawrence Wackett, a pilot, lost his life in 1944 – a loss that had a profound impact on his private life.

As the Korean War unfolded, the need for more modern aircraft became evident through combat experiences. Wackett rose to the occasion once more, taking a leading role in the development of the Sabre Jet Fighter. This aircraft became a cornerstone of Australia's Air Force. When it was time to replace the Sabre, Wackett played a central role in the design of the Dassault Mirage, marking his final major contribution before his retirement in 1966.

In recognition of his exceptional services to aviation, Lawrence Wackett was knighted in 1954 and received the prestigious Oswald Watt Gold Medal in 1974. These accolades underscored his enduring legacy and influence in the field of aviation.

Sir Lawrence Wackett, KBE, DFC, AFC, passed away on March 18, 1982.

Sir Lawrence Wackett's contributions to Australian aviation are indeed remarkable. From the trenches of World War 1 to the design rooms of jet fighters he played a pivotal role in establishing the Royal Australian Airforce.

He demonstrated ingenuity and bravery in war, received prestigious awards, and helped establish the country's aircraft industry. His legacy continues to inspire, earning him the title of 'The Father of Australian Aviation'.

Australian Air Aces

On a Wing and a Prayer

Historically '**flying ace**' was a French term used to describe one of its pilots who shot down 7 enemy aircraft in 1915. This designation later became a way to honour any pilot credited with downing enemy planes during aerial combat. Typically, a pilot needed to achieve five or more victories to officially earn the title of an ace. The actual number of victories to officially qualify as an **ace** has varied, but is usually considered to be five or more.

During WW1 Australia was the only dominion to have its independent air force with pilots serving in a range of units from the Australian Flying Corps (A.F.C), the Royal Flying Corps (R.F.C), and later, the Royal Airforce (R.A.F).

From World War I through subsequent conflicts, up to 81

Australian pilots earned the prestigious title of "air ace" while serving their nation during wartime. Among them, several remarkable individuals stand out:

First Aussie aviator VC winner

Francis Hubert (Frank) McNamara, born in 1894 was an Australian recipient of the Victoria Cross the highest decoration for valour awarded to a member of the British and Commonwealth forces.

Serving with the Australian Flying Corps, he was honoured for his actions on 20 March 1917, when he rescued a fellow pilot who had been forced down behind enemy lines. McNamara was the first Australian aviator—and the only one in World War I—to receive the Victoria Cross.

He later became a senior commander in the Royal Australian Airforce (RAAF) rising to the rank of air vice-marshal by 1942.

He held senior posts in England and Aden during World War II. Retiring from the Air Force in 1946, McNamara continued to live in Britain until his death from heart failure in 1961.

Aces High

Robert Alexander Little

Born in Victoria in 1895, he was generally regarded as the most successful Australian flying ace with an official tally of forty-seven kills.

He travelled to England in 1915 and learned to fly at his own expense before joining the Royal Naval Airforce and posted to the Western Front achieving thirty-eight victories within a year and earning the Distinguished Service Order and Bar, the Distinguished Service Cross and the French Croix de Guerre.

Rested in July 1917, he volunteered to return to the front in March 1918 and scored a further nine victories before he was killed in action on the night of 27 May, aged twenty-two.

Roderic Stanley (Stan) Dallas

Dallas born in 1891 was an Australian fighter ace of World War I. His score of aerial victories is generally regarded as the second-highest by an Australian, after Robert Little but there is considerable dispute over Dallas's exact total. Like Little, Dallas flew with British units, rather than the Australian Flying Corps. Beyond his combat record, Dallas achieved success as a squadron leader, both in the air and on the ground. He was also an influential tactician and test pilot. His service spanned almost the entirety of World War I achieving further victories before being killed in action on 1 June 1918 while on patrol in northern France. He was aged 27.

Arthur Henry Cobby

Born in 1894 Cobby was an Australian military aviator and leading ace fighter in the Australian Flying Corps with 29 victories, even though he saw active service for less than a year.

Acclaimed as a national hero, Cobby transferred to the newly formed Royal Australian Airforce (RAAF) in 1921 rising to the rank of wing commander.

He left the Permanent Air Force (PAF) in 1936 but remained in the RAAF reserve until the outbreak of WW2.

In 1943, he was awarded the George Medal for rescuing fellow survivors of an aircraft crash. He was appointed Air Officer Commanding No. 10 Operational Group the following year but was relieved of his post in the wake of the 'Morotai Mutiny' in 1945. Retiring from the Air Force in 1946, Cobby served with the Department of Civil Aviation until his death on Armistice Day in 1955.

Elwyn Roy King

King was a fighter ace during World War I achieving twenty-six victories in aerial combat, making him the fourth highest-scoring Australian pilot of the war, and second only to Harry Cobby.

Born in New South Wales, King initially saw service as a light horseman in Egypt in 1916. He transferred to the AFC as a mechanic in January 1917 and was subsequently commissioned as a pilot. Posted to No 4 Squadron, he saw action on the Western Front scoring more "kills" than most other pilots. His exploits earned him the Distinguished Flying Cross. Returning to Australia in 1919, King spent some years in civil aviation before co-founding a successful engineering business. He joined the RAAF following the outbreak of World War II and held several training commands, rising to the rank of group captain shortly before his sudden death in November 1941 at the age of forty-seven.

Cedric Howell

Cedric Howell who was born in Adelaide in 1896 enlisted in the air force in 1916 and in November of that year was accepted into the Royal Flying Corp.

Over the next 15 months, he was credited with shooting down 19 enemy aircraft and was thus awarded the Distinguished Service Order. He had previously been awarded the military cross and the distinguished flying cross.

He entered and competed in the Great Air Race in a Martinsyde A.I. Aircraft. However, he and his navigator, George Fraser, tragically died on December 10, 1919, when their plane crashed just off the coast of Corfu.

Howell's body washed ashore and was returned to Australia for burial, but unfortunately, Fraser's body was never found.

World War 2 Flying aces

This is a selection of Australian fighter aces in World War 2.

While it has already been established that an "ace" is generally considered to be any pilot who has downed five or more enemy aircraft, the term has never been officially adopted by the Royal Australian Airforce.

Accordingly, the number of victories attained by fighter pilots was not routinely publicised by the RAAF during the Second World War.

Paterson Clarence Hughes.

Born in Cooma, New South Wales in 1917, Hughes joined the Royal Australian Air as a cadet in 1936.

In Britain following the outbreak of World War II, he began flying Spitfires sharing in his unit's first aerial victory on 8 July

1940. Serving with the Royal Air Force, he was credited with as many as 17 aerial victories during the Battle of Britain. He then began scoring heavily against the Luftwaffe the following month

His tally made him the highest-scoring Australian of the battle and among the three highest-scoring Australians of the war.

Known for his practice of attacking his targets at extremely close range, Hughes is generally thought to have died after his Spitfire was struck by flying debris from a German bomber that he had just shot down. He was posthumously awarded the Distinguished Medal and was buried in England.

Keith William "Bluey" Truscott.

Born in 1916 Keith Truscott was a WW2 fighter pilot and Australian Rules footballer with the Melbourne Football Club.

After joining the Royal Australian Airforce in 1940, he became the second-highest Australian World War II ace credited with 20 confirmed victories and 5 unconfirmed victories

After he completed his flying training in Canada, Truscott served in Britain flying Spitfire fighters. He returned to Australia in early 1942 and served in New Guinea, where he fought during the climactic Battle of Milne Bay. He was killed in a joint Australian-US training exercise off the coast of Western Australia in March 1943, at the age of 26.

Adrian Philip "Tim" Goldsmith,

On 16 September 1940, Goldsmith enlisted in the Royal Australian Air Force for service during the Second World War. Accepted for pilot training, he then graduated from his course and was posted to England in June 1941 with the rank of sergeant.

In February 1942, Goldsmith was posted to Malta during a critical period in the Siege. Flying Hawker Hurricanes and Spitfires he managed to shoot down one Messerschmitt and damage another during an aerial engagement on 21 April; his plane was damaged by a cannon shell fired by one of four enemy planes.

Within the first fortnight in May, Goldsmith had shot down a further six German and Italian aircraft and was subsequently recommended for the Distinguished Flying Medal.

By the end of his tour in July 1942, Goldsmith had amassed a tally of at least 12 aircraft shot down.

Returning to England, he served as a flying instructor before proceeding home to Australia in January 1943. On 25 March 1961, Goldsmith died in a Sydney hospital after suffering peritonitis.

A Tribute to Aces High.

Since World War I brave souls have taken to the skies, not seeking glory, but defending our land and freedom. From the dusty deserts of the Great War to the far-flung islands of the Pacific in World War II, and beyond; Australia's pilots have shown unwavering courage and dedication.

They fought with valour and resolve, often under the most perilous circumstances, knowing that the skies held the key to victory.

These pilots, from the ANZACs of the Royal Flying Corps to the modern aviators of the Royal Australian Air Force, have written a story of sacrifice and honour leaving an indelible mark on history. Their legacy lives on, as a testament to the enduring spirit of Australia.

Women of Influence

Amy Johnson and Emelia Earhart.

American-born flyer Emelia Earhart and British-born Amy Johnson, are arguably among the most inspirational and influential women of the twentieth century.

Both their stories blend grit, determination, achievement, and tragedy.

Neither were Australian but both inspired Australian women at a time when many were tied by conservative traditions. They are truly worthy of being mentioned here as influencers on Australia's female flyers who reached for the skies and shattered that once impenetrable glass ceiling.

The Influence of Emelia Earhart.

Amelia Earhart, born on July 24, 1897, in the American state of Kansas, had an unlikely introduction to aviation. At the tender age of eleven, in 1908, she witnessed one of the Wright Brothers' early aeroplanes but initially harboured no interest in flying. It was only after her first plane flight, on December 28, 1920, during a visit to an air show in California with her father, that her passion for aviation was ignited. As the plane ascended a few hundred feet off the ground, Earhart later recalled, "I knew I had to fly."

Driven by her newfound passion, Earhart worked tirelessly, supplemented by financial support from her mother, to pay for flying lessons. She eventually acquired her own plane, a bright yellow aircraft she affectionately called the "Canary," and earned her pilot's license.

In 1928, she was invited to join a historic flight across the Atlantic with pilot Bill Stultz and co-pilot Slim Gordon. As the navigator, Amelia helped guide the plane named Friendship. On June 18, 1928, after twenty-one hours of flying, they landed in Wales, marking Earhart as the first woman to cross the Atlantic by air.

This achievement catapulted Earhart to hero status in the United States, with a ticker tape parade in New York City and an audience with President Calvin Coolidge at the White House. This just ignited Amelia's ambitions further and now was driven to make the same trip across the Atlantic, but this time she wanted to pilot the plane and make the flight by herself.

On May 20, 1932, she took off from Harbour Grace, Newfoundland aboard a bright red single-engine Lockheed Vega aeroplane. She intended to make the same flight that Charles Lindbergh had made five years before and fly to Paris, France.

The flight proved to be fraught with danger as extremely bad weather closed in by thick clouds, and often her windshield and wings were covered with ice. Fourteen hours later she had crossed the Atlantic Ocean but had to cut the flight short, landing in a cow pasture in Londonderry, Northern Ireland completing her 14-hour flight.

Earhart became only the second person, after Charles Lindbergh, to solo cross the Atlantic, earning her numerous accolades, including the Distinguished Flying Cross from Congress.

Over the following years, Earhart continued to break records, such as being the first person to fly solo from Hawaii to California. Her advocacy for women's rights and her passion for aviation made her a prominent figure, inspiring countless women worldwide.

However, Earhart's most audacious dream was to circumnavigate the globe, and she embarked on this historic journey on June 1, 1937, alongside navigator Fred Noonan. They left Miami, Florida, flying across Africa, and Asia, and reaching New Guinea in the South Pacific.

Tragically, on July 2nd, while attempting to reach Howland Island in the Pacific Ocean, they disappeared without a trace. There have been many theories about what happened to the flight, but no one knows and her plane has never been found.

The influence of Amy Johnston

Amy Johnson, born in England in 1903, had a different journey into aviation. She later attended the University of Sheffield – where female students were rare – but graduated with an economics degree in 1925.

After graduating with an economics degree, she initially worked as a typist in London but following a broken romance began looking for different opportunities away from the grind of her nine-to-five desk job.

Johnson's fascination with flying began when she impulsively took a bus ride to the Stag Aerodrome in North London, where she watched planes take off and land.

In September 1928, Johnson embarked on flying lessons at the London Aeroplane Club and earned her pilot's license the following year, becoming the first British woman with a ground engineer's license. She quit her secretarial job to work full-time as a mechanic at the aerodrome, demonstrating her determination to excel in the male-dominated world of aviation.

In 1929, Amy announced her bold plan to become the first woman to fly solo from England to Australia. With the support of her father and Charles Wakefield, the Lord Mayor of London and founder of Castrol Oils, she purchased a second-hand de Havilland Moth named Jason. Meticulously planning her route and refuelling stops, she embarked on her journey from Croydon, England, on May 5, 1930.

Flying approximately 18,000 kilometres with no radio over 20 days, Amy relied solely on a compass and wristwatch to guide her.

Her historic solo flight culminated in Brisbane on May 24, 1930, where thousands of well-wishers turned out to greet her and hear her radio address to the nation, saying *'Hello, Australia. I am very tired and I'm going to bed. I hope to see you in a few days.'* And she did. From the moment she touched down, Australia immediately embraced her. She received overnight celebrity status normally reserved for film stars with thousands attending her many engagements across the country.

And the feeling was mutual.

'How vastly different are the conditions in Australia from what I had pictured them! The Australians have swept me off my feet...'

The period following the First World War was an exciting era for all aviators. They became the rock stars of their time and their exploits were often the subject of popular songs. Amy was no exception, her journey to Australia inspired many songs by famous writers of the time, including Jack O'Hagan and Jack Lumsdaine.

That's why the tragedy of Amy Johnson that unfolded so shocked the nation.

January 1941 Amy disappeared while flying from Blackpool to the RAF base Kidlington. It is believed Amy Johnson hit adverse weather conditions or ran out of fuel and was forced to bail out as her plane went down in the Thames Estuary.

The crew of a passing ship HMS Haslemere saw her descending by parachute and hitting the water. Despite the terrible weather conditions, including heavy snow and a strong tide, the commander of the Haslemere jumped in to save her but unfortunately, he failed, and Johnson's body was never found.

Despite the tragic endings to both their stories, the legacies of Emelia Earhart and Amy Johnson endure. Their pioneering spirit and determination inspired countless women in Australia to break through those conservative social barriers at that time. Two trailblazers, with unyielding courage, left an indelible mark on Australian aviation history and continued to inspire countless women to this day.

First Female Flyers

Hilda Hope McMaugh: first Australian woman to fly in the UK

In the history of aviation, the achievements of women have often been overshadowed by their male counterparts. However, if one delves deep into the vaults and archives of national and state libraries, there lies a treasure trove of stories—stories of remarkable women who defied conventions and soared to new heights in the early 20th century. Among them is the distinguished Australian nurse, Hilda Hope McMaugh.

Hilda McMaugh's story begins in the aftermath of World

War I, a time when many progressive women, who had served their nations during the war, sought new challenges beyond the battlefield. Hilda had spent her wartime years in France, working primarily for the Red Cross. But as the armistice marked the end of her military service, she returned to England, her home since leaving her birthplace of Uralla in New South Wales.

The year was 1919, and it was the beginning of the golden age of aviation. Dramatic aerial feats, round-the-world flights, air races and barnstorming were capturing the imagination of a world emerging from war.

Although young Nancy Bird Walton was just four years old and long before the exploits of Amy Johnston and Amelia Earhart, flying became a sudden passion for Hilda McMaugh. Then, in a remarkable feat, after just one month of training, she successfully passed her tests and was awarded the Royal Aero Club's Pilot Certificate. She accomplished this feat by flying a Centaur 4 machine. The London's Sunday Times reported on November 16, "*She is the first Australian woman to take the certificate and the first woman in England to pass since the Armistice. But she may never fly again.*" A fleeting passion it would seem.

Two months later, Australian newspapers who had followed her pursuits reported that Miss McMaugh had no plans to continue flying Down Under, but she would return to Australia to continue her nursing duties. Hilda returned to significantly contribute to her local community at Uralla.

Today her legacy continues long after her passing in 1981. She left enough funds for the establishment of the McMaugh Gardens Retirement Home. Hilda Hope McMaugh may not have pursued a prolonged flying career, but at least she had achieved her goal of being "No.1" on the list of Australian women pilots.

Florence Taylor – First woman to fly in Australia

Florence Taylor is another name that shines brightly in the early history of Australian aviation. Born in 1879, Florence was a trailblazer in multiple fields—she became Australia's first qualified architect, engineer, editor, businesswoman, feminist, and newspaper columnist. However, it was her passion for flying that set her apart.

On December 5, 1909, Florence Taylor made history by launching herself airborne from the Narrabeen sandhills near Sydney. She did this in a glider built by her husband, George Taylor, a pupil of Australia's pioneer of flight, Lawrence Hargrave. George had already made a series of flights in his biplane glider, including the longest at 110 yards (just over 100 meters), marking the first successful heavier-than-air flight in Australia.

Reporting on Taylor's triumphant flight, a *Sydney Morning Herald* reporter wrote:

'At "let go" the wind immediately lifted the machine to the full

length of the guide ropes and dragged the operators so fast that two let go; the machine now soared towards the ocean, and at the water's edge the remaining guide ropes were loosened, the machine made a huge leap upwards. Mr Taylor by careful manoeuvring kept the machine well under control.'

So excited by the flight and ease of control, he invited his wife to try.

The *Sydney Morning Herald* continued to report on Florence's flight, stating that when she released the glider, it was immediately lifted by the wind, and Mr Taylor managed to keep the machine well under control. Florence, who shared her husband's interest in aviation, eagerly took to flying with enthusiasm.

On that day, December 5, 1909, she became the first woman in Australia to fly a heavier-than-air machine, albeit a glider. Remarkably, it would take another 18 years before an Australian woman would obtain an official pilot's license.

Millicent Bryant

March 28, 1927, Millicent Bryant achieved a significant milestone in Australian aviation history. At the age of 49, and just nine months after the death of her husband, Millicent passed her private pilot's license test at the Australian Aero Club of New South Wales. Flying a De Havilland DH60 Moth aircraft, she earned her pilot's license, making her the first woman in Australia to do so. Her accomplishment was a remarkable feat, given the conservative norms and challenges of that era.

Millicent didn't stop at obtaining her pilot's license; she also participated in the inaugural Ladies Oakes flying race on October 6, 1927, coming in second place. Margaret Reardon claimed the first position, with Evelyn Follett securing third place in the three-entrant race.

However, Millicent's flying career was tragically short-lived.

Just ten months after gaining her pilot's license, she met a tragic end in a ferry accident on Sydney Harbour.

The deadliest incident in the harbour's history took place when the passenger liner *Tahiti* ran over and sank the small ferry on 3 November 1927.

Her funeral was marked by a touching tribute, as five planes flew over the ceremony and dropped a flower wreath in her honour. Her leather flying helmet, a poignant reminder of her pioneering spirit, found its place in the National Library of Australia before being moved to the Powerhouse Museum in Sydney in 2005.

Nancy Bird-Walton, O.B.E (1915-2009)

Nancy Bird-Walton, O.B.E., is a name that has left an indelible mark on Australian aviation.

Born on October 16, 1915, she would become a pioneering Australian aviatrix, known affectionately as "The Angel of the Outback."

Nancy's journey into aviation began at the tender age of four when she attempted to launch herself off the backyard fence after hearing news about the Great England-Australia Race, which was captivating the nation.

By the time Nancy turned 13, she experienced her first joy flight in a Gipsy Moth at a local fair. This momentous occasion ignited a lifelong passion for flying, prompting her to immediately begin saving for flying lessons.

Her actions, though unapproved by her father, led her to cross paths with the legendary aviator Charles Kingsford Smith, who would play a pivotal role in shaping her aviation career.

It was Kingsford Smith who, upon recognising Nancy's enthusiasm, invited her to come to Sydney for flying lessons.

To Nancy Bird-Walton, Charles Kingsford Smith was the epitome of aviation greatness. She described his flying technique as *"one of beauty and precision, exactly timed, judged, and balanced."*

In the beginning, Kingsford Smith may not have taken her seriously, given Nancy's petite stature, standing at just five feet tall (150 cm). However, her determination soon earned her his respect.

At the age of 17, in 1933, Nancy embarked on her journey to become a pilot—a time when women pilots were a rare breed. In the early 1930s, it was not common for women to even wear long pants, let alone fly an aircraft.

Despite these challenges, Nancy Bird-Walton displayed unwavering persistence and continued her chosen career path.

In 1935, she was hired to operate an air ambulance service in outback New South Wales, known as the Far West Children's Health Scheme. Nancy used her own Gipsy Moth aircraft as an air ambulance, saving countless lives by flying patients to hospitals. This was no small feat, as the airstrips available in those days were often unmade and riddled with hazards like rabbit holes. Yet, Nancy never once crashed a plane, a testament to her skill and courage. Flying in those early days was fraught with risk, as navigation instruments were basic, and road maps were often used instead of aviation maps.

In the 1950s, Nancy Bird-Walton founded the first Australian Women Pilots' Association, paving the way for countless women to enter the world of aviation. She broke barriers and stereotypes by becoming the first woman to enter the 'Powder Puff Derby', an all-women race in America. While her domestic life included marriage to Englishman Charles Walton and the birth of two children, Anne Marie and John, her passion for flying never waned.

In 1935, Nancy Bird Walton achieved another milestone when she became the first woman in Australia to obtain a license allowing her to carry passengers. This license remained with her for most of her life. Her remarkable journey continued until her passing in 2009 at the age of 93.

Thanks to Nancy Bird Walton's pioneering spirit, women in Australia today serve as airline captains, helicopter musterers, search and rescue pilots, and even flying nuns.

While Nancy Bird Walton wasn't the first Australian woman to obtain a pilot's license—that distinction belongs to Millicent Bryant in 1927—she was the first to obtain a license allowing her to carry passengers, an accomplishment that symbolised her enduring influence on Australian aviation.

Maude Rose (Lores) Bonney, AM, MBE

Brisbane-based aviatrix Lores Bonney once fumbled for a revolver she kept as she sat slumped anėd distressed in the wreck of her half-submerged upturned Gypsy Moth. But the weapon had been lost in the crash and may not have been used anyway as she survived.

She once landed her damaged aircraft in Perth to discover a wing was holding it together by mere threads. She dodged death dozens of times in her solo record-breaking career and thankfully kept meticulous notes to record it all.

Lores Bonny was born Maude Rose Rubens on 20 November 1897 in Pretoria, South Africa but even as a toddler was well-travelled by the time her family settled in Melbourne, Australia.

It was while helping the War effort in 1917 that she met and fell in love with Harry Bonner and moved to Brisbane where she met intrepid airman Bert Hinkler who took her on her first flight which would prove to be a life-changing experience.

Since her very first joy flight with Hinkler, Bonny acquired as much information from every pilot she encountered even asking Charles Kingsford Smith for advice but was unimpressed, she noted, 'with his condescending manner'.

Within months her devoted husband Harry bought her a gipsy-moth biplane which she named 'My Little Ship' and wasting no time immediately embarked on her first record-breaking career.

The first of Bonney's four major solo flights took place on Boxing Day 1931. Leaving Brisbane at 4.30 a.m., she reached Wangaratta, Victoria, at 7.20 p.m., in time for dinner with her father.

Between 15 August and 27 September 1932, she became the first woman pilot to circumnavigate Australia.

Aiming to be the first woman to fly from Australia to England, Bonney left Brisbane on 10 April 1933.

She was caught in a tropical storm en route and was forced to land on a beach in Thailand. But as she approached, a herd of water buffalo walked into her path, and she crashed-landed into the nearby sea. It was this distressing incident that she felt for her revolver. She managed to free herself and arrange for her plane to be shipped to India for repairs.

On 25 May she resumed her flight and on 21 June landed at Croydon, England.

Despite the credit she had long craved, Bonny wasn't satisfied and longed to set more records.

On 9 April 1937 at the age of 39, Bonney took off from Brisbane crossing the Indian Ocean to become the first person to fly from Australia to South Africa arriving at Cape Town on 18 August.

It was on the route she heard that Amelia Earhart was hot on her tail and hoped they would meet in Khartoum. However, they missed each other by one day and less than a month later, on 2 July 1937, Amelia Earhart disappeared over the Pacific Ocean.

Returning to Brisbane, Bonny considered other destinations however with Earhart's disappearance her husband opposed them all.

Unlike so many of her peers Bonny lived to the age of 96 and 57 years after setting her last record she died in a Queensland nursing home on 24 February 1994.

She may be regarded by some as the forgotten aviatrix but I hope this helps make amends, lest we forget too.

These extraordinary women—Hilda Hope McMaugh, Florence Taylor, Millicent Bryant, Nancy Bird-Walton, and Lores Bonney—have left indelible marks in the history of Australian aviation. Their unwavering passion and courage have inspired generations of aviators, both men and women.

They may have faced turbulent headwinds, but they soared above them all, leaving behind a legacy of achievement and inspiration for all who dared to dream of flight. And let's not forget the others who followed.

Less We Forget

Tributes in Brief

As we have witnessed, the history of flight is rich with remarkable men and women who pioneered the skies for others to follow. Their dedication and innovation have undoubtedly left a lasting impact on how we travel and explore the world to this day.

They may not have all received the same widespread acclaim, but their contributions are integral to the history of aviation.

* * *

Jessie Maude "Chubby" Miller

Born in 1910, Jessie Maude "Chubby" Miller embarked on an incredible journey that would etch her name in aviation history. In June 1927, she departed the shores of the United Kingdom alongside Bill Lancaster, aboard an AVRO Avian aircraft. After 159 days filled with challenges and triumphs, she achieved an unprecedented milestone – becoming the first woman to fly solo from Australia to Britain.

Florence "Bobby" Mary Terry

Florence "Bobby" Mary Terry was destined to soar. Born in 1898, she made history in 1929 as the first Australian woman to own an aeroplane. Her ambitions knew no bounds; she obtained a commercial pilot's license, specializing in seaplanes. Notably, she joined a group of aviators who flew in formation with the legendary Amy Johnson during her 1930 arrival in Sydney, marking another achievement in her illustrious career.

Margaret "Meg" Skelton

Margaret Skelton's aviation journey was marked by resilience. As one of six women pilots escorting Amy Johnson during her historic flight into Sydney in 1930, Skelton showcased her unwavering commitment to flying.

Although she faced obstacles during the Great Depression and did not own an aircraft, she earned her pilot's license in 1929.

Sadly, by 1953, her flying days had come to an end.

Phyllis Rogers Arnott

Phyllis Rogers Arnott, born in Sydney in 1907, hailed from Arnott's biscuit manufacturing family.

Her pursuit of flight began in earnest when she became the first Australian woman to obtain a commercial pilot's license 'B' in 1931. Motivated by her brothers' flying pursuits, Phyllis continued to take to the skies. Eventually, she shifted her focus to studying engines and left flying behind. In a tribute to her legacy, Qantas honoured her by naming an A380 aircraft after her.

Freda Thompson

Freda Thompson's ascent to the skies was also fuelled by passion and determination. Born in Yarra, Victoria, in 1909, she took her first flying lesson in 1930 at Essendon. Hailing from a wealthy family, Thompson bartered music lessons for flight lessons and earned her private 'A' pilot's license in 1929. In 1932, she achieved a commercial 'B' pilot's license, making her the fifth Australian woman to do so. Her remarkable journey included a trip to England to acquire a De Havilland Moth Major plane. Although she missed the Great Race, she was celebrated upon her return to Australia as the first woman to fly from England to Australia (reversing the achievement of Maud Bonney). Later, she served as the commandant of the Woman's Air Training Corps in Victoria and continued her flying adventures, amassing a collection of forty-seven trophies before finally retiring in 1980.

Margaret "Peggy" McKillop Kelman OBE

Peggy McKillop, born in Scotland in 1909, embarked on her flying journey in 1931 at the Aero Club of NSW. With determination and dedication, she earned her pilot's 'A' license in 1932 and a commercial pilot's license 'B' in 1935. Her first paid flying job was for Nancy-Bird Walton in 1935. During an 11-week tour of country New South Wales, fate intervened as she crossed paths with a young grazier who owned his own plane. Colin Kelman's journey with Peggy took them to London in 1936, where they were married. In an adventurous twist, they purchased a second-hand twin-engined plane and decided to fly back home to Australia.

After her husband's passing in 1964, Peggy's involvement in the Australian Women's Pilot Association deepened.

She became the Australian head of the International Women Pilots' Association and received the OBE in 1978 for her outstanding contributions to women's aviation.

Ivy Pearce Hassard

At the age of 16, Ivy Pearce Hassard embarked on her flying odyssey.

After taking flying lessons, her father gifted her a Tiger Moth aircraft. In 1935, she earned her pilot's 'A' license, marking the beginning of a successful flying career. Her list of achievements included winning numerous races, mastering aerobatics, and earning the title of the fastest female pilot in the 1936 Adelaide Centenary Air Race. By 1950, Ivy had transitioned away from flying and embarked on a successful career as a fashion industry icon on the Gold Coast, Queensland.

Amy Gwendoline Stark Caldwell OBE

Bondi-born Gwen Stark's aviation story unfolded in 1939 when she obtained her pilot's license. She later became one of the first women appointed to a position in the Women's Australian Auxiliary Air Force. Before the outbreak of World War II, she actively participated in the Australian Women's Flying Club, which eventually evolved into the New South Wales branch of the Women's Air Training Corps. By 1940, Amy assumed the role of commandant. Her post-war adventures took her to Europe, where she played a role in the Berlin Air Lift at a Royal Air Force station in Germany. In 1964, she assumed the position of federal president of the Australian Women's Pilots' Association and was bestowed with the Order of the British Empire in 1968 in recognition of her outstanding contributions to aviation.

Nancy Ellis Leebold MBE

South Australian native Nancy Ellis embarked on her flying journey at the Royal Aero Club in Mascot. In 1942, she achieved her private pilot's 'A' license, followed by her commercial license 'B' in 1946. Nancy's remarkable achievements included becoming the first Australian woman to fly heavy aircraft and serve as the First Officer of a Lockheed Lodestar for Air Cargo Pty Ltd in 1950. She was also among the 35 charter members of the Australian Women Pilots' Association in 1950.

A pivotal moment in her career came in 1952 when she piloted a Lockheed Shooting Star T33A, making her the first Australian woman to fly a jet aircraft. Her exceptional accomplishments continued as she received an Amelia Earhart Memorial Scholarship in 1954, studying in the United States and England. In a remarkable feat, Nancy flew a single-engine Miles Messenger back to Australia with her husband as a passenger. Her journey was crowned with the Evelyn Follett award from the Australian Women Pilots' Association for the year's most outstanding flying performance over 12,000 kilometres.

Evelyn Follett

Evelyn Follet's story is a testament to tenacity. In August 1927 she became the third Australian woman to earn a pilot's licence.

Despite her petite stature, she overcame obstacles to pursue her passion, even using a cushion to reach the controls. Together with her brother, Captain Follett, and Captain Bunny Hammond, she co-founded an airways administration company called Adastra. Evelyn was one of the five female pilots who flew in formation with Amy Johnson during her historic arrival in Sydney during her England-Australia flight in 1930.

May Bradford Shepherd

May Bradford Shepherd's journey into aviation was ignited by an overheard conversation about aviator Amy Johnson. Her competitive spirit led her to take flying lessons in 1931, where she achieved a remarkable feat: she became the only woman in Australia to hold both a first-class pilot's 'A' and 'B' license, in addition to 'A' and 'B' electrical ground engineers' certificates. In 1936, she joined the ranks of female pilots, including Nancy Bird, Lores Bonney, Freda Thompson, and Ivy Pearce, to complete the Adelaide Centenary Air Race, flying a Klemm Eagle aircraft she had built in the workshops at Mascot. Tragically, her promising career met a devastating end in January 1937 when her plane, on take-off at Mascot, clipped another, resulting in a fiery crash that claimed her life along with her two passengers.

* * *

These extraordinary women of aviation, against all odds, defied gravity and conservative norms to achieve greatness in the skies. Their stories have continued to inspire generations of women to reach for the stars and shatter glass ceilings in the world of aviation.

The Flying Kangaroo

Flying Aces Hudson Fysh KBE DFC (1895-1974) – Paul McGinness DFC DCM (1896 – 1952)

The motives of young Australians who enlisted during the First World War were multifaceted, encompassing patriotism, a thirst for adventure, and an escape from domestic hardships. Their journey often began in the Middle East, before the fateful Gallipoli campaign and the harrowing experiences on the front lines of France. Among the survivors, some faced traumatic physical and mental injuries, others returned to resume their lives; yet, as we have frequently discovered, some took to the skies to return as heroes.

The post-war years of the early 20s saw many Air Force servicemen returning home, equipped with flying experience and visions of how aviation could link the vast expanses of the Australian continent.

This period laid the foundation for the advancement of commercial aviation in Australia and the establishment of what would become one of the oldest airlines in the world. A co-founder of Qantas also helped propel another icon into our skies – The Royal Australian Flying Doctor.

Hudson Fysh

Initially, Hudson Fysh served as a ground soldier in Gallipoli later writing:

"We lived like rats in holes and hung on to our Hillside with the enemy on one side and the beach, a few hundred yards away, on the other".

In July 1917, Fysh was transferred to the Royal Flying Corps flying alongside fellow Australian Paul McGinness.

In August 1918, he attacked and destroyed two hostile enemy aircraft, while on another occasion he engaged in combat, destroying three more.

On January 8, 1919, he was awarded the Distinguished Flying Cross. His award read: "For gallantry in air combat and in attacking ground objectives".

Paul McGinness

Paul McGinness was born on February 14th, 1896 in Warrnambool, Victoria. He was not only the co-founder of Qantas with Hudson Fysh but also a highly decorated pilot in both WWI and WWII.

His biography tells of a daring 'jump-out-of-the-trenches-and-charge-at-the-enemy' episode. He was the sole survivor of a group of 150 men attributed to the fact that he was knocked unconscious early in the attack and believed killed. He was retrieved for medical treatment after dark.

McGinness was awarded the Distinguished Flying Cross and the Distinguished Conduct Medal. He also received the title of Air Ace being credited with 7 aerial victories.

However, he is now best remembered for his role as co-founder of Qantas.

Qantas

On returning home Fysh and McGinness were determined to keep alive their passion for flying and their belief in the future of aviation in Australia. They accepted an offer from the defence department to survey an air race route from Longreach in Queensland to Katherine in the Northern Territory. By land, this journey took them 51 days but left the pair more convinced than ever that an air service was needed to link remote regions of the vast continent.

They began by raising funds to start a company, initially offering an air taxi service as well as leisure and sightseeing flights.

In November 1920 Queensland and Northern Territory Aerial Services Limited (soon abbreviated to QANTAS) was formed.

Their first aircraft were two flimsy biplanes based at their headquarters in Winton before moving to Longreach a few years later. Financier Fergus McMaster then joined the two pilots as chairman but did not stay long, while Hudson Fysh remained with Qantas until 1966.

The first move to a regular airline service came in November 1922 with a government contract to operate a mail service between Charleville and Cloncurry. Passenger services soon followed with the introduction of a four-seater de Havilland.

In 1929 the route network was extended to finally reach Brisbane.

In 1927, Hudson's friendship with a like-minded devout Christian the Reverend John Flynn led to the signing of a joint vision of a flying doctor service for people living in remote parts of Australia.

Hudson proposed the use of his QANTAS' single-engine De Havilland; a refurbished war plane named *Victory* fitted to carry a stretcher, doctor and pilot.

In 1934 overseas expansion began when Qantas and Imperial Airways (a predecessor of British Airways) jointly formed Qantas Empire Airways Limited (QEA). This name lasted until 1967 when the airline was renamed Qantas Airways, as it remains today.

They also introduced the first Empire flying boat service to meet a growing demand.

Following the outbreak of the Second World War, Qantas played a pivotal role in the war effort. Flying unarmed planes through war zones and at times under enemy fire, the airline supplied the front-line troops, evacuated the wounded and undertook surprising escapes. But not without considerable loss. Between 1942 and 1943, Qantas lost eight aircraft against the Japanese with the deaths of over sixty passengers and crew. The Qantas hangar and flying boat service facilities were destroyed in the first Japanese air raid on Darwin on 19 February 1942. A Qantas Empire flying boat narrowly escaped destruction during the raid.

Following the war, the Australian aviation industry rebounded when QEA was nationalised as the government bought the shares of both BOAC and Qantas. Service continued in partnership with BOAC to the UK, but was also expanded significantly into SE Asia.

In 1950 the Flying boat service was introduced to several overseas Pacific destinations.

In 1956, the airline entered the jet age as propeller aircraft were phased out. It coincided with the year Qantas drew worldwide attention to flying visitors (and the flame) to Australia for the Melbourne Olympics.

By 1959 the airline had become the first, outside the United States, to use Boeing jets and in 1979 introduced the world's first business-class seating.

By now tens of thousands of post-war migrants had caught their first glimpses of their new homeland on board the 'flying kangaroo', so named after Qantas's original route to England with its multiple 'hopping stops'.

Since then, the flying kangaroo has become the iconic Australian emblem.

Major changes would continue through the late 1990s.

Following the collapse of competitor Ansett Australia in 2001, Qantas enjoyed around a 90% share of the domestic Australian market. This would reduce the later growth of Virgin Blue.

In 2003, Qantas launched its low-price subsidiary Jetstar Airways.

* * *

For almost 100 years, through times of war and peace, national celebration and national service; the flying kangaroo has embodied the spirit of Australia.

Whilst this brief history perhaps does not do justice to the Australian icon, it hopefully serves as a reminder of its humble beginnings, and those visionaries who shared the dreams that flight would benefit all.

The Flying Doctor

Inspirational Pioneers

"If you start something worthwhile – nothing can stop it."
Founder, the Very Reverend John Flynn.

* * *

'Romantic Tales of Outback Australia' created the vision that would drive young John Flynn to become another of Australia's most inspirational aviation pioneers.

Born at Moliagul in central Victoria in 1880 he moved several times after the death of his mother until finally settling with his father in the western suburb of Sunshine.

However, the romantic outback tales soon began to merge with stories of hardship and struggles when his father's business partners mounted an unsuccessful commercial venture in the far north of rugged Australia.

These conflicting tales only fuelled Flynn's desire to discover more about life in the outback.

Meanwhile, John Flynn continued his schooling in Melbourne until 1898 when he accepted 'the call' to join the ministry.

Initially, he financed his studies working at Church Home Missionary Centres around Victoria then in 1907 commenced a four-year course in divinity at The Melbourne of University. He graduated in 1910, was ordained as a Minister of the Presbyterian Church one year later and headed off to spread his missionary message throughout rural and remote areas of Victoria and South Australia.

It wasn't until early 1911 that John Flynn discovered 'the road to the real outback and saw first-hand the rigours of outback life and the lack of medical care available to inland settlers'.

After a year of discovering life in the outback Flynn was commissioned to prepare a report and returned to present his findings to Church leaders in Melbourne.

Flynn's report prompted the General Assembly to act upon his recommendations appointing him as the head of the Australian Inland Mission (AIM).

Then in 1917, Flynn received an inspirational letter from former Melbourne medical student and aviator Lieutenant Clifford Peel.

The young pilot and war hero suggested 'the use of aviation to bring medical help to the outback'. However, Clifford Peel would never realise his letter would become the blueprint for the creation of the Flying Doctor Service. He died a short time later after being shot down in France at the age of just 24.

For the next ten years, Flynn stepped up his campaign for an aerial medical service for inaccessible rural areas of Australia.

His dream was to provide a 'mantle of safety' for the people of the bush but it would take another Melbourne supporter before his vision became a reality.

Hugh Victor McKay was a Melbourne industrialist who had invented a machine which was revolutionising wheat harvesting around the world.

Together with his father and brother John, they built a prototype of a combine harvester in a factory workshop in the Melbourne suburb of 'Braybrook Junction'.

In 1885 Hugh McKay patented this invention which had been developed in what was for many years the largest factory in Australia.

Residents even voted to change the name of their suburb to Sunshine, a name that today reflects the enormous success of the 'Sunshine Harvester'. It was another lost irony as Sunshine was also the Melbourne home of John Flynn.

McKay left his Sunshine home in 1922 when he achieved his long-held ambition of owning Rupertswood at Sunbury.

HV McKay, as he was known, was not only an inventor and entrepreneur; the son of Irish protestants also left a large bequest for 'an aerial experiment' which enabled John Flynn to achieve his long-held ambition to get his Flying Doctor Service airborne.

It was around this time, that Flynn also met Hudson Fysh, a founder of QANTAS.

In 1927 QANTAS and the Aerial Medical Service signed an agreement to operate an aerial ambulance from Cloncurry, Queensland.

When the first pilot took off on 17 May 1928 he was flying a single engine, timber and fabric bi-plane named 'Victory'.

The pilot, Arthur Affleck, had 'no navigational aids, no radio and only a compass'. He navigated by landmarks such as 'fences, rivers, river beds, dirt roads or just wheel tracks and telegraph lines'.

In its inaugural year, the Aerial Medical Service (which changed its name to the Flying Doctor Service in 1942 and the Royal Flying Doctor Service in 1955) flew 50 flights to 26 destinations and treated 225 patients.

Flynn's dream had now become a reality.

'The Flying Doctor Service had a doctor, a pilot and John Flynn, the man with the vision, but they lacked the communication technology to deliver services efficiently'.

Victorian-born Alfred Traeger helped to hurdle this barrier with the invention of a pedal-operated generator to power a radio receiver.

By 1929 people living in isolation were able to call on the Flying Doctor to assist them in an emergency.

Using the Flying Doctor Service network, the School of the Air was established in Alice Springs in 1951 also 'providing children living in remote areas the opportunity to interact with their teacher and other students' supplementing their correspondence lessons.

In May 1950, Flynn attended what was to be his last Council meeting; he died of cancer exactly one year later.

Flynn was cremated and his ashes placed at rest under the Flynn Memorial, just west of Alice Springs at Mount Gillen – the very centre of the vast territory to which he brought communication, medical comfort and pastoral care.

Those currently on board the Flying Doctor proudly boast 'it's a service that provides extensive primary health care and 24-hour emergency service to people all around the nation' thanks to those visionaries who were truly Australian pioneers.

The Ansett Story

Reg Ansett

Much has been written about Reg Ansett and not all has been favourable. But like all pioneers of aviation, we cannot overlook his enormous achievements.

Reginald Myles Ansett: A Pioneer of Australian Aviation

The story of Reginald Myles Ansett is a testament to the power of determination, innovation, and entrepreneurial spirit. While opinions about him vary, there is no denying the profound impact he had on the aviation industry in Australia.

Born in 1909 in a small town in Victoria, Reg Ansett's journey from humble beginnings to the founding and expansion of Ansett Australia is a remarkable tale of a man who dared to dream big and turn his dreams into reality.

Ansett's Early Years and Entrepreneurial Spirit

Reg Ansett's early life was marked by modest means and limited formal education. He left school at the young age of 14 but showed an innate aptitude for engineering. He attended Swinburne Technical College to qualify as a knitting machine and sewing machine mechanic.

In 1929, at the age of 20, Ansett made a pivotal decision that would alter the course of his life. He cashed in a life assurance policy and used the funds to embark on a new adventure – learning to fly. He enrolled in flying lessons, unaware he was setting out on his path of destiny.

A Year of Adventure in the Northern Territory

Before fully immersing himself in aviation, Ansett embarked on a journey that would add diversity to his life experiences. He took a gap year of sorts, travelling to the Northern Territory to work as an axeman with a survey party. This venture exposed him to the vast and rugged Australian outback, providing valuable life lessons and shaping his character.

The Brief Peanut Farming Episode

Ansett's adventures did not stop there. During his time in the Northern Territory, he briefly contemplated becoming a peanut farmer. However, the solitude and isolation that accompanied this endeavour ultimately convinced him that was not his future. Ansett was a man of action and engagement, and the prospect of a solitary life on a peanut farm did not align with his spirit.

Entrepreneurial Ingenuity: The Car Delivery Service

With a meagre savings of £70, Ansett returned to Victoria, where his entrepreneurial spirit truly began to shine. He purchased a second-hand Studebaker car, and in a move that foreshadowed his future business acumen, he converted the car into a vehicle for transporting passengers and freight between Maryborough and Ballarat. This venture marked his official entry into the world of business, albeit as a young and relatively inexperienced entrepreneur.

The Shift to Hamilton and Business Success

The initial car delivery service, while promising, proved to be financially challenging. Ansett, undeterred by setbacks, decided to relocate his base of operations to Hamilton in December 1931. This move proved to be a pivotal one, as he established a more successful car delivery service route between Hamilton and Ballarat.

However, his success did not go unnoticed or unchallenged. In 1932, Robert Menzies, who would later become Prime Minister of Australia, introduced legislation aimed at protecting the state-run Victorian railways from competitors like Ansett, who were siphoning off both passenger and freight revenue. Ansett, always quick to adapt and innovate, responded with audacity.

The Birth of Ansett Airways Pty Ltd

In 1935, Ansett took a bold step by purchasing a six-seat Fokker Universal aircraft and registering Ansett Airways Pty Ltd. This marked the official establishment of an air service between Hamilton and Melbourne, a move that signalled his entry into the aviation industry.

To subsidise this fledgling airline service, Ansett offered "joyrides", thrilling passengers with acrobatic displays and providing flying lessons. These joyrides not only generated additional income but also captured the imagination of the public, laying the foundation for Ansett's reputation as an aviation entrepreneur.

Challenges and Competition in the Aviation Industry

During this period, the Australian domestic airline travel sector was largely dominated by Australian National Airways (ANA), established in 1936. However, the government, under the Chifley Labor administration, sought to establish a state-owned alternative, leading to the creation of Trans Australia Airlines (TAA).

The 1950s marked a significant turning point in Ansett's career. With Robert Menzies as Prime Minister, Ansett resolved to challenge the dominance of ANA, which was considered the weaker of the two major airlines. He embarked on a strategic plan to revamp his airline, purchasing new aircraft and gradually upgrading cabin service. Most notably, he implemented fare discounts that hit ANA hard, rendering them unable to compete effectively.

The Takeover of ANA and Further Expansion

The turning point came in January 1957, when ANA's founder, Sir Ivan Holyman, unexpectedly passed away. Ansett seized the opportunity and orchestrated a takeover of ANA, a move that solidified his position in the aviation industry.

Throughout the 1960s and 1970s, Ansett's entrepreneurial spirit continued to drive his empire's expansion. His interests extended beyond aviation, leading him to invest in television stations (namely Channel 0/10) in Melbourne and Brisbane, among other major businesses. However, this diversification also made him vulnerable to takeovers, setting the stage for future challenges.

Challenges and Controversies

The 1970s brought a fair share of challenges for Ansett. In March 1975, air hostesses went on strike, creating significant disruptions in his airline's operations. During this period, Ansett made a regrettable remark, referring to the striking hostesses as a "batch of old boilers". This incident underscored the evolving rights of workers in the aviation industry.

Ansett also faced controversy regarding his opposition to the recruitment of women as pilots. However, this stance was challenged in the High Court in 1979, which ultimately ordered him to employ Deborah Wardley as a trainee pilot (see later chapter). This marked a significant moment in the history of gender equality in the aviation sector.

The Takeover and the End of an Era

Late in 1979, still reeling from the High Court decision and facing mounting challenges, Ansett faced another significant development. Sir Peter Abeles and media mogul Rupert Murdoch launched a successful takeover of Ansett Transport Industries and appointed Reg Ansett as chairman.

This marked the end of an era for the man who had played such a pivotal role in the Australian aviation industry.

By 1980, Ansett's aviation operations were spun off into a separate entity, Ansett Air Freight, further reshaping his business landscape.

Sadly, on September 14, 2001, due to financial difficulties, Ansett Australia entered voluntary administration, leading to the cessation of all group operations.

Legacy and Final Years

Reg Ansett's entrepreneurial spirit, bold vision, and determination were undeniable. He overcame numerous odds, challenged industry giants, and left an indelible mark on Australian aviation.

Reginald Myles Ansett passed away at his home in Mt Eliza on 12 December 1981, leaving behind a legacy that extended far beyond his lifetime. Like many complex figures in history, Ansett's legacy has not all been remembered favourably however it's essential to acknowledge his immense accomplishments.

Deborah Wardley

Ansett Nemesis

Until the year 1927, Australia upheld a stringent and unjust practice – barring women from obtaining commercial pilot licenses or pursuing commercial aviation careers within its borders. This policy stemmed from deeply ingrained conservative beliefs that insisted *a woman's place was in the home.* These attitudes had long curtailed women's aspirations and ambitions – and not just in aviation.

However, in the backdrop of this prevailing discrimination, there were global inspirations. Remarkable female pilots like

Amelia Earhart and Amy Johnston had already ignited the imaginations of Australian women, encouraging them to break free from these constrictions.

On March 28, 1927, a significant milestone was achieved when Millicent Bryant became the first Australian woman to secure a pilot's license but with restrictions on carrying passengers. Tragically, her promising journey was cut short later that year when she lost her life in a terrible ferry accident in Sydney Harbour.

Another source of inspiration emerged in the form of Nancy Bird, who achieved the historic feat of becoming the first female pilot in the Commonwealth authorised to carry passengers. But not in Australia.

Nevertheless, it would take a courageous and determined Melbourne woman to shatter the glass ceiling and usher in a new era for Australian female pilots.

In the year 1979, I first met Deborah Wardley as a journalist covering her written complaint to Victoria's Equal Opportunity Board, accusing Ansett Airlines of gender-based discrimination in their hiring practices.

Deborah Wardley's qualifications were beyond reproach. She had held a full pilot's license since the age of 18, accumulating an impressive 2600 flying hours. Moreover, she possessed a background as a trained high school mathematics and science teacher. Her determination to qualify as an airline pilot was evident as she diligently sent application letters to Ansett Airlines for a period of two years. Yet, despite her undeniable qualifications and relentless pursuit, she claimed that the airline had consistently bypassed her while readily hiring ten of her male counterparts who were also flying instructors.

Wardley's battle against this gender discrimination was a gruelling and protracted one. In response to her accusations,

Ansett Airlines brazenly countered with a shocking claim that a woman's menstrual cycle rendered her unfit for the role of a pilot. Such an argument underscored the deeply ingrained gender bias of the era.

Reg Ansett, the founder of the airline, vociferously denied any form of discrimination. However, he unabashedly expressed his conviction that women were inherently unsuited for the role of airline pilots. This stance was legally challenged and galvanised a public response that would eventually rock the foundations of Ansett Airlines.

Public demonstrations and a call for a boycott of the Airline ensued, resulting in a substantial loss of business for the company, estimated to be as high as 50 per cent. To justify their stance, Ansett Airlines put forth a series of contentious arguments:

1. They asserted that pilots needed physical strength, despite the absence of any established strength test for pilots.
2. Ansett contended that women's menstrual cycles posed an insurmountable obstacle to their suitability as pilots.
3. The airline argued that the potential for pregnancy and childbirth would disrupt a woman's career to the extent that it would compromise safety and impose additional costs on the company.

The tide finally began to turn when the Victorian Equal Opportunity Board delivered a resounding verdict – Ansett's continued refusal to employ Deborah Wardley, based on her gender was illegal. The board awarded her $14,500 in damages and issued a compelling order – Ansett must include her in their next pilot training program.

However, the battle was far from over. Ansett Airlines

proceeded to appeal the decision, initiating a complex legal saga. The company delayed its pilot training intake, and the case was subsequently brought before the Supreme Court, where, much to Ansett's chagrin, their appeal was dismissed.

Undeterred, Ansett Airlines elevated the dispute to the High Court. Remarkably, during this legal battle, the airline employed Deborah Wardley, albeit with considerable pressure from labour unions. Yet, despite these developments, when Deborah completed her classroom training in December 1979, she was still not assigned to flight training.

At this crucial juncture, as Ansett Airlines deliberated on their next move, fate intervened in the form of Rupert Murdoch and Peter Ables, who launched a takeover bid of Ansett Airlines. This momentous event marked the beginning of the end for Ansett and heralded a new chapter in the life of Deborah Wardley.

In early January 1980, Wardley leveraged a unique connection. She had previously served as the flight instructor for John Calvert-Jones, Rupert Murdoch's brother-in-law. Following Murdoch's successful takeover of Ansett Wardley sought his assistance, and her plea did not fall on deaf ears.

In a swift and decisive move, Murdoch issued a memo directing that Deborah Wardley be treated on par with any male trainee. This directive allowed her to commence flight training without further delay.

On January 22, 1980, a historic moment unfolded as Deborah Wardley piloted her first commercial flight as a co-pilot, from Alice Springs to Darwin. Her perseverance and the unwavering public support she received finally shattered the gender barrier in Australian aviation.

1993 Deborah Wardley embarked on a new phase of her flying career. She assumed the role of a pilot for KLM, taking charge of international flights. Her ascent in the world of aviation was nothing short of remarkable but not without determination.

By the year 2007, she had ascended to the position of senior captain of the Airbus A380 with KLM. But there remained a vestige of gender discrimination.

One particular incident during her tenure with KLM is a testament to her resilience and wit. Upon discovering that a woman was piloting his plane, a passenger expressed concerns.

Wardley, in her characteristic composed manner, approached the passenger who immediately recognised her Australian accent. She inquired if the passenger had previously flown with her and arrived safely at their destination. The passenger confirmed a safe arrival, prompting Wardley to quip, "Well, what are you complaining about, then?" Her response was a fitting retort to a fading public gender bias.

Upon reaching the mandatory retirement age for airline pilots in Europe, Deborah Wardley returned to Australia in 2008. She joined Jetstar as a Safety Investigations Manager and occasionally took to the skies as a pilot to maintain her Airbus A320 rating.

In July 2012, Wardley embarked on another remarkable chapter in her journey. She joined Tiger Air Australia as an Airbus A320 Captain and Instructor, effectively accomplishing her mission of breaking down any stubborn gender barrier.

In November 2023 Deborah (Wardley) Lawrie, aged 70, was flying as a captain for Virgin Airlines.

To honour her achievements as a trailblazer for women in aviation the NSW government announced a bridge to Sydney Airport would be named in her honour.

Responding with humility and humour she noted that 'The Deborah Wardley Flyover', part of the Sydney Gateway project, would stand above the Sir Reginald Ansett Drive, named after the man who once refused to hire her.

Aminta Hennessy and Ray Clamback

Modern Aviation Pioneers

In 1978 Aminta Hennessy became the first Australian woman to fly solo across the Atlantic in a single-engine aircraft. One would think by 1978 there would be few records still to break but she then became the first woman to have completed solo back-to-back flights across the Pacific.

As we have discovered through the determination of women like Deborah Wardly; Aminta Hennessy is another of those trailblazers, who defied the odds and etched her name in our aviation history.

Born in England, Aminta's fascination with aviation was sparked by the legendary Australian aviation pioneer, Nancy Bird Walton. Bird's achievements inspired Aminta to follow her dreams and embark on a lifelong journey in the cockpit. Little did she know that her path would be paved with challenges, but she was determined to conquer them all.

Aminta's aviation career began with an unexpected twist. She initially served as a patron of the Royal Flying Doctors Service, a vital organisation that provides medical care to remote Australian communities. However, her life took a dramatic turn when she decided to take flying lessons in Alice Springs. This choice led her down a new trajectory, one that would see her become a pilot herself.

To fund her flight training, Aminta worked tirelessly, pulling double shifts at a bar during the afternoon and evening, earning her the affectionate moniker "the flying barmaid of the north".

Despite her dedication and newfound passion for flying, the 1960s and '70s were a time when not everyone was comfortable with a woman in the pilot's seat.

Recalling her struggles, Aminta shared a poignant encounter with an airline captain who callously dismissed her ambitions, saying, "*Over my dead body, I don't want any women in my cockpit.*"

It was in these moments that she found her unwavering partner in life and aviation, Ray Clamback, who became her staunchest supporter.

In 1978, the pair made history together when they opened a flying school at Bankstown Airport in Sydney. Their shared love for flying led them on an extraordinary journey.

Aminta and Ray both ventured to America, each purchasing a Cherokee Warrior aircraft. To bring their prized possessions back to Australia, they faced the challenge of flying across the Atlantic

Ocean. In doing so, Aminta Hennessy became the first Australian woman to achieve a solo Atlantic crossing.

The flying school they established thrived for more than four decades, leaving an indelible mark on countless aspiring pilots. However, as the years passed, time took its toll. Ray faced two harrowing crashes in the Pacific, with Aminta witnessing the last one from above.

On October 5, 2004, Ray, Aminta Hennessy, and senior flying instructor Lyn Gray embarked on a flight from Hilo, Hawaii, to American Samoa.

During the journey, Ray's Cessna 182 suddenly lost oil pressure, forcing him into an emergency ditching in the Pacific Ocean. Although Ray managed to escape the sinking aircraft, it was a heart-stopping ordeal for Aminta and the others circling above who were unaware Ray had even survived for there were no visible signs of life from the scene below.

The situation seemed dire, but Ray's tenacity had prevailed and was rescued by the U.S. Coast Guard after spending six and a half hours in the Pacific waters.

Ray's remarkable survival was not an isolated incident. In 1999, he and his co-pilot found themselves adrift in the Pacific for ten hours when their plane's engine failed shortly after leaving Los Angeles. These experiences made Ray contemplate his future, but he returned to their Sydney flying school just one day after arriving home.

At 67 years old, Ray acknowledged, *"I've been doing it for a long time, but now I am a lot weaker."*

Meanwhile, Aminta also felt the weight of time on her shoulders and joints, making it increasingly challenging to enter and exit an aircraft.

'I can't think of anything I've achieved more than what I ever dreamed of'.

'I can't think of anything that I wished I had done, I have done it all, but now it's just too difficult and dangerous so I am stopping'.

Aminta Hennessy's legacy in aviation extended far beyond her historic solo flights across the Atlantic and Pacific. She was a co-owner of Clamback & Hennessy, the longest-established flying training and charter organisation at Bankstown Airport in Sydney. Additionally, she played a pivotal role in establishing the Australia Association of Flight Instructors and served as its president for more than five years.

Amintas's commitment to aviation safety and education was evident in her creation of 'Fear of Flying' clinics in collaboration with the Australian Women Pilots' Association and Qantas Airlines in 1979. For a decade, she dedicated herself to running these clinics, helping individuals conquer their anxieties about flying.

Furthermore, her lasting impact on aviation was marked by the establishment of the Australian Association of Flight Instructors, advancing the profession and ensuring future aviators received the best and safest instructions.

Aminta Hennessy's journey is just another example of women breaking down barriers in aviation.

The closing of Clamback & Hennessy on June 30th, 2019, marked the end of an era in Australian aviation.

However, Aminta's legacy lives on in the hearts of all those she inspired and the aviation institutions she helped establish.

David Warren:
Black Box Inventor

In the early hours of January 10, 1954, at Rome's Ciampino airport, a historic moment was unfolding. It was just after 9:30 a.m. when the world's first commercial jetliner taxied for take-off on the last leg of its journey from Singapore to London. The De Havilland DH 106 Comet, a marvel of modern aviation, smoothly accelerated down the runway, destined for a cruising altitude of 35,000 feet and a speed of 460 miles per hour. This was more than 100 miles per hour faster than the fastest propeller-driven airliner of the time.

Twenty minutes into the flight, as the Comet reached an altitude of 27,000 feet, a cryptic message was sent from the aircraft to another BOAC flight. 'George How Jig from George Yoke'. Peter, did you get my..." The message abruptly broke off at 9:51 a.m.

Shortly after this incomplete message, shocking news began to emerge. The Comet jet airliner had crashed into the Mediterranean, and all 35 people on board were missing, feared dead. Giovanni di Marco, a fisherman who had been in the vicinity, recounted what he had witnessed. He heard three explosions in quick succession, and then, several miles away, he saw a silver object flash out of the clouds, trailing smoke before it plunged into the sea. There was a massive cloud of water, and when Giovanni reached the scene, all was eerily still. He and others began the grim task of recovering bodies, but there was little else they could do.

Tragically, this was the third crash involving a Comet aircraft since its introduction into service on May 2, 1952. The most devastating of these accidents occurred on the first anniversary of the jet's debut when all 43 people on board perished shortly after take-off from Calcutta, India. An inquiry into this calamity attributed it to an unusually severe storm, which led to structural failure and a fire onboard, ultimately resulting in the crash.

In the wake of these incidents, the entire Comet fleet was grounded, pending further investigations.

Despite the grounding, Comet flights eventually resumed on March 23, 1954. However, a mere two weeks later, on April 8, another Comet, this time on a charter flight from Rome to Cairo, met a similar fate, crashing into the Mediterranean shortly after take-off. This string of disasters raised alarm bells and cast a shadow of uncertainty over the future of jet travel.

British aviation was in dire need of answers to these baffling and tragic incidents. They turned to an unexpected source for help: Melbourne, Australia, in 1954.

At the heart of this inquiry was the Defence Science and Technology Group, or DST as it was known, an organisation responsible for providing science, technology, and aeronautical expertise to meet Australia's defence and national security needs. The DST's Research Laboratory at Fishermen's Bend in Melbourne had earned a stellar reputation on the world stage. As the senior research scientist at this prestigious institution, Dr David Warren found himself at the centre of a pivotal moment in aviation history.

Dr Warren's connection to aviation disasters was deeply poignant. He had lost his father in a 1934 plane crash in Bass Strait. The thought often haunted him – what if there had been a way to record the events leading up to that fateful crash?

At the time, Dr Warren's primary research focus was on fuels for new gas turbines entering aeronautical service. He was asked to investigate whether fuel tanks might be exploding on the Comet aircraft, potentially contributing to the crashes. As he delved deeper into the details of these accidents, a revolutionary idea began to take shape in his mind. What if, he wondered, there was a way to record everything that happened in the cockpit of an aircraft, right up to the very moment of a crash? What if the pilots themselves could tell us what was happening?

Dr Warren recognised the limitations of traditional tape-recording technology – it would not survive a burning plane crash. However, he remembered seeing a miniature wire recorder at a recent trade show, an instrument designed for recording music. In a moment of inspiration, he realised that a similar recorder could be carried in the cockpit of an aircraft to capture critical information.

This revelation marked the birth of what we now know as the "black box". Dr Warren, in collaboration with his instrument colleagues at the Aeronautical Research Laboratory (ARL), embarked on the challenging task of developing a device capable of recording both flight data and voice recordings onto a wire. This innovation represented a significant leap forward in aviation technology.

However, as with many ground-breaking concepts, particularly in Australia at the time, Dr Warren encountered resistance in getting his idea off the ground. His efforts to garner interest domestically yielded little result. Frustrated but undeterred, he drew upon his early experience as a teacher, recalling the effectiveness of "show and tell" over merely explaining ideas.

Dr Warren decided to take a proactive approach. He set out to build a demonstration recorder that could continuously store up to four hours of speech, including crucial flight instrument readings. Still, there was minimal interest from Australian authorities. It seemed as though his ground-breaking idea might remain unacknowledged.

The breakthrough moment arrived unexpectedly during an informal visit to ARL by former British Air Vice-Marshal, Sir Robert Hardingham, in 1958. Over a casual lunchtime conversation, Sir Robert asked Dr Warren to demonstrate his "unofficial project". The demonstration left a profound impression on Sir Robert, who immediately recognised its potential.

Eager to see this innovation put into practice, Sir Robert sent both Dr Warren and his invention to London. There, they received a far more encouraging reception from the Ministry of Aviation, which announced plans for the installation of black box flight recorders in British aircraft. This development spurred other countries, including Canada and the United States, to follow suit.

However, the story in Australia took a different turn. It wasn't until a tragic event in 1960 that the nation began to fully appreciate the importance of Dr Warren's invention.

A Fokker Friendship aircraft crashed in Queensland, leading the inquiry judge to strongly recommend the installation of cockpit voice recorders in all airliners. Australia then became the first country in the world to make the cockpit voice recorder mandatory.

Since that pivotal moment, the "black box", which is not black but orange for easy identification, has become a standard fixture in all aircraft worldwide. It has proven invaluable in unravelling the mysteries of countless air disasters, shedding light on the events leading up to accidents, and subsequently contributing to significant improvements in aircraft safety on a global scale.

The Comet crashes, which initially baffled investigators, were eventually traced back to a fatigue crack at a square window corner near the radio direction-finding aerial in the roof of the aircraft. This seemingly minor structural flaw led to catastrophic consequences, tearing the plane apart in mid-air, an event the pilot would not have even seen.

In recognition of his pioneering efforts, Dr David Warren was posthumously honoured with the most prestigious award in the aviation industry in 2016. His outstanding contribution to transport safety had not only significantly improved aviation but also served as a blueprint for safety measures in other modes of transportation, such as rail and shipping.

He passed away in 2010 at the age of 85 and was buried in a casket bearing the inscription: Do not open.

Dick Smith

Aviation Adventurer to Philanthropic Pioneer

My first encounter with the remarkable Dick Smith occurred in October 1976 just 4 years after he obtained his flying licence. We were part of a re-run of the historically famous Perth to Sydney air race. As a journalist with the Seven Network, I was among the several hundred competitors flagged away from Jandakot airport by aviation legend Sir Douglas Bader. Our first destination was an isolated runway at Forrest on the Nullarbor.

The significance of Forrest in aviation history stretches back to 1928 when the Australian Government called tenders for a Transcontinental air mail service between Adelaide and Perth. Forrest had already emerged as a crucial rail stopover point, featuring formal fine dining with silver service and entertainment in a large hangar. Housekeeping staff were employed to operate a hostel where they prepared evening meals and breakfast. Forrest was not so much a town, as a siding and fettlers camp on the Transcontinental Railway line.

For participants in the Perth to Sydney air race, our first evening meal at Forrest was a simple barbecue, followed by a night spent beneath the wings of our planes under the starry Outback sky. As dawn broke, it was met with a thunderous chorus of outrage and obscenities from numerous competing pilots.

The source of this collective fury was none other than Dick Smith himself. During the night, accusations had arisen that he had affixed stickers to their aircraft to promote his burgeoning electronics business.

The issue wasn't the stickers *per se* but the damaging aftermath when pilots attempted to remove them, inadvertently peeling off the aircraft's enamel paint. The damage to their aircraft was far from minimal.

It was a rare instance where Dick Smith found himself at the receiving end of abusive words, but this incident also hinted at the unyielding entrepreneurial spirit that would later define him. This spirit culminated in the establishment of Dick Smith Electronics, a company that, by 1982, had become synonymous with small electronic items, from calculators to computers.

Beyond electronics, 'Dick Smith Foods' emerged as a vehicle to champion Australian-made products, thereby supporting local farmers.

He wasn't content with just business success; Smith extended his influence to the realm of media production, founding Smith and Nasht, a company dedicated to producing films to raise awareness about climate change and other pressing issues.

Then came a pivotal transition in his life. Dick Smith sold his businesses, embarking on a new career, one steeped in philanthropy and continuing his spirit of adventure. The extent of his charitable contributions is staggering; he once claimed to have had no idea how much he gave away, with estimates soaring as high as $500 million.

In the realm of aviation, Dick Smith remained an indomitable force. On February 13, 1977, he conceived and initiated airline flights over Antarctica. Smith chartered a jumbo jet to take paying passengers on a 10-hour, 5,000-kilometre flight over the Red Centre in pursuit of Lasseter's lost reef. Once again, profits from his adventures were directed toward charitable endeavours.

But it was his audacious attempts that garnered the most attention. In October 1991, Smith became only the second person to fly over Mount Everest, a feat that cemented his reputation as an intrepid explorer. He also left his mark through well-publicised practical jokes, such as the 1978 April Fool's Day stunt in which he was expected to tow an iceberg from Antarctica into Sydney Harbour, ostensibly to provide a new source of fresh water.

Yet, it was his achievements as an adventurer that truly set him apart. From his initial expedition as a 20-year-old to climb the forbidding Balls Pyramid rock, he went on to claim numerous world firsts:

- First solo trans-Atlantic flight by helicopter (1982).
- First solo circumnavigation by helicopter (1983).
- The first round-the-world aeroplane flight, landing at both the North and South poles (1989).
- First non-stop crossing of Australia by balloon (1993).

- First east-west circumnavigation by helicopter (1995).
- The first trans-Tasman balloon crossing (2001).

These accomplishments firmly established Dick Smith as one of Australia's most daring adventurers. His contributions extended beyond the realm of exploration, earning him recognition as Australian of the Year in 1986.

He also served as the head of the Civil Aviation Authority between 1990 and 1992, and again from 1997 to 1999. In 2015, he received the prestigious Companion of the Order of Australia.

Dick Smith's life journey, from the controversy of adhesive stickers on aeroplanes to becoming a celebrated adventurer and philanthropist, is just another example of the power of determination and the impact one individual can have on their country and the world.

In 1986 he was selected as Australian of the Year. He served as head of the Civil Aviation Authority between 1990–92 and 1997–99 and in 2015, he was awarded the Companion of the Order of Australia.

He still maintains a streak of controversy, campaigning for change to air traffic control regulations, a fair-go for all Australians and a cut to immigration.

However, it's his indelible mark on the history of aviation, business, and philanthropy in Australia that will be his lasting legacy.

Richard de Crespigny

Modern Aviation Hero.

Not since the Bali bombings in 2002 or the downing of Malaysian Airlines Flight 17 in 2014, had we faced a story that potentially threatened the lives of so many Australians.

A personal perspective:

It was the 4th of November 2010.

As a news presenter on Channel Ten Melbourne, I had just completed a live midday update on stories of no particular consequence.

Minutes later, I re-entered the newsroom amid a non-urgent murmur of conversation, an occasional phone ringing and a background of low-level sound from monitors attached to the newsroom walls.

Only when a voice suddenly called out, "Possible plane crash" did the atmosphere dramatically change. The alert was attributed to an unconfirmed social media report on Twitter. Rarely did we ever react to such vague or unreliable sources but calls were made to our media contacts at Moorabbin and Tullamarine airports; just in case. Someone else switched two of the monitors to CNN and Fox News. Again nothing!

Then a short time later Fox News repeated the same social media report. But now their story was accompanied by a disturbing account from the Indonesian island of Batam.

Locals villagers on the ground were describing debris falling from the sky and striking a school. Most disturbing were scenes of dragging what appeared to be part of an engine cowling. Not just damage from any plane but clearly marked by the Qantas red and white logo.

In newsroom terms, all hell broke loose as our emergency news handbook was immediately implemented.

All available resources were thrown into the unfolding crisis which paled into insignificance compared to the crisis unfolding on the flight deck of a giant A380 Airbus 7000 feet above Batam.

* * * *

Melbourne-born and educated Richard de Crespigny received his first taste of a flying career as a 14-year-old when his father organised a tour of the RAAF Academy at Point Cook in Victoria. It was a defining moment in his young life planting seeds of a dream to become an RAAF Officer and Pilot. The rest is history.

In 1975, aged 17, he joined the RAAF. One year later, he began flying and remained with the RAAF until 1986 when he joined Qantas. Beginning with Boeing 747s he converted to the A380 in 2008 to become one of Qantas' most senior captains.

At 10 am on the morning of 4th of November 2010 Captain de Crespigny, the pilot of Qantas flight 32 was awaiting final clearance for take-off from Changi's Singapore airport.

He had already reported all was normal on board the world's largest and most advanced commercial jetliner carrying 420 passengers and a crew of 29. He was about to begin the final leg of an international flight from London to Sydney.

In an unusual twist, there were five pilots in the cockpit on this flight. In addition to the normal crew of pilot-in-command and co-pilot, there was a relief pilot and two additional check captains. That experience would prove to be most fortuitous.

Singapore take-off:

The take-off, he would later reflect, was perfect. Captain Richard De Crespigny pushed the thrust levers forward from their idle position. The four Rolls-Royce Trent 900 engines roared to life as 14 litres of jet fuel and 120 tonnes of air poured into them every second. The 464-tonne aircraft surged down the runway and launched into the air at 350km/h before its 22 wheels retracted. It was 9.57 a.m. (Singapore time) and everything was running exactly to plan as they went through the routine 'After Take-off' checklist.

As they passed 6000 feet De Crespigny would reveal 'everything was looking like a picture-book day'. It was only as they reached 7400 feet at 10.01 and were about to turn off the seatbelt sign when they heard a relatively small boom and a slight jolt. Then just one second later, there was a huge BOOM described as 'nothing they had ever heard before'.

Damage:

Four minutes into the flight, the A380 suffered a catastrophic engine failure. Shrapnel from the number 2 engine punctured part of the wing, damaging the fuel system and causing leaks. These leaks caused a fuel tank fire. Hundreds of pieces of shrapnel blasted through the engine, travelling at more than 2.6 times the speed of sound. They would later describe 'debris flying with such a velocity in all directions [that] you could not see the individual pieces'. The sound was compared to that of marbles rattling against corrugated iron plates as the shrapnel penetrated the wing and fuselage.

Of the aircraft's 22 different systems, 21 were damaged. In all, 650 wires and network cables were severed. Only half of the electrics and hydraulics were operational. The aircraft then responded by flying out of balance. They had also lost breaking and flap control with its landing gear so damaged it could only be lowered using an emergency gravity option. Fuel and hydraulic fluid were now leaking from the badly damaged left wing.

The second explosion had blown the engine apart damaging the wing, by then the cockpit crew were struck by a cascading sound of alarm bells, blaring horns and whooping sirens as they threw themselves into crisis mode. Amid the crisis, a pungent smell from sparks and fumes permeated the cockpit as a flash fire ignited inside the wing fuel tank."

Response

CAPTAIN Richard de Crespigny a veteran of 35 years with Qantas and the Royal Australian Air Force, appeared unruffled but fully in control.

Throughout the emergency de Crespigny never forgot he was in charge of a team and by sheer good luck, he had five experienced captains in the cockpit with him. They all began processing emergency procedures but to no avail. They divided tasks, made calculations and offered suggestions, although, in the end, it was the captain's call.

De Crespigny was an aviation junkie who lived and breathed flight technology. He knew so much about the A380 that he had utter faith in the machine he was flying. As long as he could keep it flying.

Cabin reaction:

The antithesis of the cockpit crisis was the silence among passengers in the cabin. The first explosion was audible to some but there was no doubt there was a problem after the second blast which shook the entire aircraft. Some looked out the window and could see the damaged wing and leaking fuel. Gradually the passengers recovered from the moment of shock and began showing signs of concern as cabin staff began a series of reassurances.

Flight deck:

Back on the flight deck, it was decided to delete all written procedures from their computers and fly by gut instinct alone. As De Crespigny later wrote. 'At least the plane was still flying and I had full respect the aircraft could be safely landed. Just how that was to be achieved was the main problem'.

At least one hydraulic system was down and the anti-lock braking system was causing the No. 1 and No. 4 engines to go into a "degraded" mode. The landing flaps had been severely damaged but the aircraft could still be landed manually by a gravity drop emergency system.

It took 50 minutes to complete their final assessment. They could not accurately calculate a landing distance as the plane was still fully laden with fuel or whether the strip could cope with their landing speed.

After stabilising the aircraft and virtually bringing it under their control, the crew then decided to fly a holding pattern close to Singapore's Changi Airport as all ground emergency procedures went into action.

'We had no option but to come in fast knowing the brakes would malfunction because we were loaded with excess fuel which we could not jettison'.

'With a broken wing, little roll control, no autopilot or auto-thrust and these added problems, we calculated that we would probably stop just 139m short of the end of the 4km-long runway'.

The landing:

Just before commencing the landing procedure he followed all his Air Force training and conducted what they called control checks — manual tests of various crucial controls they had learnt in a dress rehearsal for such a crisis. It would not normally be done in an aircraft with passengers aboard, but they were a long way from normal.

As they descended below 1000 feet the flight warning

computers were activated blaring 'SPEED! SPEED!', something they had only ever experienced in a simulator exercise. More shockingly, moments before touchdown, they heard the one warning no pilot ever wants to hear, 'STALL! STALL!'. The warning bells and ECAM checklists kept coming simultaneously as the control tower was instructing them to shut down the three remaining engines.

As if things could not get any worse, two more electrical generators failed, taking out most of the remaining systems. Then more computers also failed, cockpit and cabin lights went off, emergency lights illuminated, and bells and alarms were sounding throughout the cabin. Evacuation messages began flashing up on the screens in front of passengers. Despite this, the cabin crew still managed to maintain control and attempted to keep passengers calm.

Finally, the aircraft at 50 tonnes overweight, made its emergency landing on the longest runway into Changi Airport, stopping just 100 metres from the end of the strip, 35 knots faster than normal and blowing four tyres in the process.

New crisis:

But the crisis was far from over as they faced the enormous risk of fire. The brakes were white-hot and ready to ignite the leaking aviation fuel flooding the area around them. Evacuations down emergency slides are difficult and dangerous at the best of times but with the risk of fire erupting on the ground, a difficult decision was made to keep passengers and crew on board until it was safe.

The Changi fire controller then informed the crew that one engine had failed to respond and was still running. They could do nothing until fire crews hosed down the brakes and covered the leaking fuel with foam.

Once conditions were safe Captain De Crespigny gave the order to disembark on the opposite side of the aircraft and onto buses. Three-and-a-half hours after landing, the last passengers were safely on their way to the terminal. Ten minutes later, De Crespigny arrived and personally shook their hands, disclosed all the details of the flight and offered care and reassurance to everyone who had been on board.

The aftermath:

By five o'clock in Melbourne, 'Ten News First at Five' was the first bulletin to broadcast details of the 'Miracle of Flight 32' including the heroics of their captain Richard De Crespigny. It was also noted that *'perhaps the spirit of aviation pioneer Nancy Bird Walton, after whom the A380 had been named, was flying with them throughout their ordeal'.*

More details would follow in the days ahead which would change the face of aviation and its safety and maintenance procedures.

Rolls Royce would later determine the direct cause of the fire and resulting engine failure was due to a misaligned counter bore within a stub oil pipe leading to fatigue fracture. It was assured the problem would never happen again.

As a result of their near-death experience, many passengers and crew suffered psychological injury and launched a class action against Rolls Royce which was resolved by way of settlement.

Almost on the eve of the 10th anniversary of averting what could have been one of the world's worst air disasters, De Crespigny announced he was taking early retirement. He may have saved

440 lives but not even Qantas could save his 45-year professional flying career.

In March 2020 he announced his retirement. "*I am currently stood down and am in limbo, not having flown since March I will take early retirement effective November 30.*

"I'll look up to A380s overhead with memories similar to those of Neil Armstrong when he looked to the sky and saw his footprint on the moon."

In 2016 he was awarded the Order of Australia for his contributions to aviation safety.

Today he has become one of Australia's most sought-after corporate speakers, with presentations focusing on crisis management, leadership, problem solving and teamwork.

Richard de Crespigny proved to be a living example of an aviation hero and a credit to many of his predecessors and their amazing flying machines who set such high standards from the very beginning of Australia's proud aviation history.

Space

Aussie Astronauts

Long before man's first flight; the stars in the heavens acted as shining beacons for those exploring the boundaries of our planet Earth.

As such, the story of flight would not be complete without acknowledging those Australians who have since achieved the ultimate flight out to those heavens above; as astronauts.

Australia's first astronaut Dr Paul Scully-Power.

Born in Sydney, in 1944 at a very young age Scully-Power developed a passion for the sea after riding a balsa wood surfboard imported from Hawaii.

In January 1967, after graduating from the University of Sydney, Scully-Power was approached by the Royal Australian Navy to set up their first oceanographic group.

In March 1974 as an Australian Navy Exchange Scientist, he left Australia and worked for the U.S. Naval Underwater Systems Centre, in New York, London Connecticut and Washington DC.

During this period, he was invited to assist the Earth Observations team on the Skylab Project where he worked in space oceanography for each successive manned spacecraft mission.

On the 5th of October 1984, the 40-year-old Australian was invited to join the shuttle orbiter Challenger's 13th mission as chief scientist in oceanography.

It was a mission of firsts in many ways – the first space shuttle mission to carry a crew of seven, the first to have two women on board and Paul Scully-Power, about to become the first Australian-born astronaut in space.

He was also the first astronaut with a beard raising fears it may have restricted his space helmet. It obviously didn't.

His seat on the shuttle was not that of a lifelong dream, but rather the culmination of a career that saw him move from oceans to space. "Going to space was just one of those things that happened," he says. "I loved surfing, so I figured I'd study Oceanography". "When you look at Earth from space and realise how much water surrounds the planet, it all makes sense'. Observing the earth's oceans from space, he discovered ocean spiral eddies, which changed how scientists now predict weather and global warming.

During his mission, which lasted for 8 days, he conducted various scientific experiments and operated the shuttle's robotic arm. By the time he'd finished his flight, he had completed 133 orbits of the Earth and logged 8 days, 5 hours, 23 minutes in space.

After his historic flight, Scully-Power continued to push the boundaries in many fields since that historic space mission. He co-founded the Ripper Group, establishing himself and his company as leaders in AI drone technology.

In recognition of his contributions to science and exploration, Scully-Power was awarded several honours, including the NASA Distinguished Service Medal, the Order of Australia, and the U.S. Navy's Meritorious Service Medal.

At the time of researching this tribute in 2023, Paul Scully-Power at 79 years of age was alive and well in Sydney.

Andrew 'Andy' Thomas

Australian Astronaut

Unlike Australian Astronaut Paul Scully-Power, Andy Thomas was a young boy who grew up with a fascination for rockets and a dream of one day becoming an astronaut. It was a dream he pursued relentlessly, ultimately becoming Australia's first member of NASA's elite astronaut corps. Over 12 years, Andy embarked on four space missions, accumulating a total of 6 months in space.

His journey first led him to study mechanical engineering at the University of Adelaide. After completing his studies, he began his career in aerodynamics research at Lockheed Aeronautical Systems Company in Georgia, USA. Remarkably, he climbed the ranks and became the manager of its Flight Sciences Division at the age of 35.

Andy Thomas applied for the astronaut program and at the age of 40, he was selected to undergo NASA's gruelling 12-month training, which included experiencing up to 40 bouts of 'weightlessness' each day, all in pursuit of his dream of becoming a fully-fledged member of the astronaut corps

Finally, in May 1996, Thomas achieved the position of payload commander in the six-person crew of the space shuttle Endeavour on a 10-day mission. This marked the historic moment an Australian had been chosen as a NASA astronaut (excluding Australian-born Paul Scully-Power who was a civilian specialist).

Following this milestone, Thomas relocated to Russia for a year of intensive preparation for a mission that included spending 141 days aboard the Mir space station. As the last NASA astronaut to participate in the Shuttle-Mir Science Program, he needed to learn the Russian language and become proficient in Russian systems and technology.

In 2001, Thomas embarked on his third space mission, joining the crew of the shuttle Discovery on a journey to the International Space Station (ISS).

His tasks on this mission included a remarkable 6-hour spacewalk to address an electrical problem on the Discovery while it was docked at the station, 320 kilometres above Earth.

Thomas took with him special Australian artefacts into space during this mission, including Sir Charles Kingsford Smith's watch and a piece of wood from Mawson's Hut in Antarctica.

Upon his return, Andy Thomas was appointed as the deputy

chief of the NASA Astronaut Office. His final space mission took place in 2005 when he returned to the International Space Station.

After a dedicated tenure spanning 22 years with NASA, Andy Thomas retired from his illustrious career in 2014, leaving an ultimate legacy in the field of space exploration and famous Aussie Flyers.

Katherine Pennell-Pegg

Australia's First Female Astronaut

In 2024, Katherine Bennell-Pegg became Australia's first qualified female astronaut. Born in 1984, Katherine grew up in the Northern Beaches area of Sydney and remembers when asked by her teacher to name three three career options she only wrote one word – 'Astronaut'.

It wasn't until 2023 that she finally achieved her childhood dream when she began training with the European Space Agency in Germany.

Only one in ten astronauts are female but the alpha-male

stereotype is quickly evolving. It may be some time before Katherine takes that giant leap for mankind but no one is more qualified and ready for blast-off than Katherine Bennell-Pegg Australia's first official female astronaut.

Box Kites to Boeings and Beyond.

A promising future

From its humble origins with 19th-century balloons to the Hargraves Box Kite breakthrough, Australian Aviation has unfolded into a saga of innovation and achievement.

The 1910-powered flight near Melbourne marked a transformative milestone, while the exploits of Australian pilots during World War I, exemplified by the Smith brothers' record-breaking 'England-Australia' journey, demonstrated unprecedented pilot skills to the world.

Qantas, founded in 1920, propelled Australia firmly onto the international stage, eventually becoming the world's longest-continually serving airline. World War II showcased the nation's pilots and spurred significant technological advancements.

The 1950s ushered in the Jet Age, with Qantas becoming the first airline outside the United States with an entire fleet of Boeing aircraft.

To the ultimate flight in outer space by Aussie astronauts.

Fuelled by innovation and a commitment to progress, Australia's pioneers have earned a reputation as world leaders in aviation. And the future too is full of promise.

Today, as the industry focuses on supersonic sustainable flying, emissions reduction, and safety, Australia is poised to continue its well-earned legacy by providing an influential role in shaping the future of global aviation.

Thanks to the following sources:

Balloons – Sydney Archives, Dictionary of Sydney, State Library of Victoria, Trove]

Airships – State Library Victoria, Tim Callanan ABC, Trove, The Argus and Airminded.com

Hargraves and Houdini – Aviation Historical Society Australia, Monash edu. au/Hargrave/Houdini. Argus Newspaper, Herald Sun, State Library of Victoria and Biography Australia.

Billy Hart / Smiths / Hinkler – Australian Aviation Hall of Fame, The Australian Dictionary of Biography, Australian National University.

Charles Ulm – The Australian Museum, National Library of Australia, author researcher Rick Searle

Gordon Taylor – National Library of Australia, Australian Dictionary of Biography, Encyclopedia of Australian Science and Wikipedia.

Sir Hubert Wilkins – State Library Victoria, the Australian War Memorial, the Australian Antarctic Division, the Australian Museum, and Author John Grierson.

Sir Lawrence Wackett – Australian Aviation Hall of Fame, Australian Dictionary of Biography, National Library of Australia.

Women at War – Australian Dictionary of Biography, Lores Bonny – The Forgotten Aviatrix, Australian Geographic.

Qantas – Founders Museum. National Museum of Australia.

Flying Doctor – Records from Royal Flying Doctor Service.

Ansett – Australian Dictionary of Biography, Sydney Morning Herald, Victorian State Library, Network Ten

Debora Wardley – Extracts from Mal Walden's 'Good News' a Brolga Publication

Aminta Hennessy – Sydney Morning Herald, Clamback and Hennessy, The Australian.

David Warren – 'Good News' a Brolga publication, Department of Defence DST, Biography Dr David Warren.

Dick Smith – Britannica, Wikipedia.

De Crespigny – Extracts from FLY! Life lessons from the cockpit of QF32 by Richard de Crespigny. The State Library of Victoria, Aviation Historical Society Australia, Sam Chui Aviation and travel. The Long Read. News.com.au

Aussie Astronauts – Biographer Colin Burgess, University of NSW, Encyclopedia Astronautica.

ALSO BY MAL WALDEN

Be Published

Publish through a successful publisher.

Brolga Publishing is represented through:

- National book trade distribution, including sales, marketing & distribution through Simon & Schuster.
- International book trade distribution to:
    - The United Kingdom
    - Sales representation in South East Asia
- Worldwide e-Book distribution

For details and enquiries, contact: Brolga Publishing Pty Ltd
ABN 46 063 962 443
PO Box 452
Torquay Victoria 3228
Australia

markzocchi@brolgapublishing.com.au
(Email for a catalogue request)